HUMAN

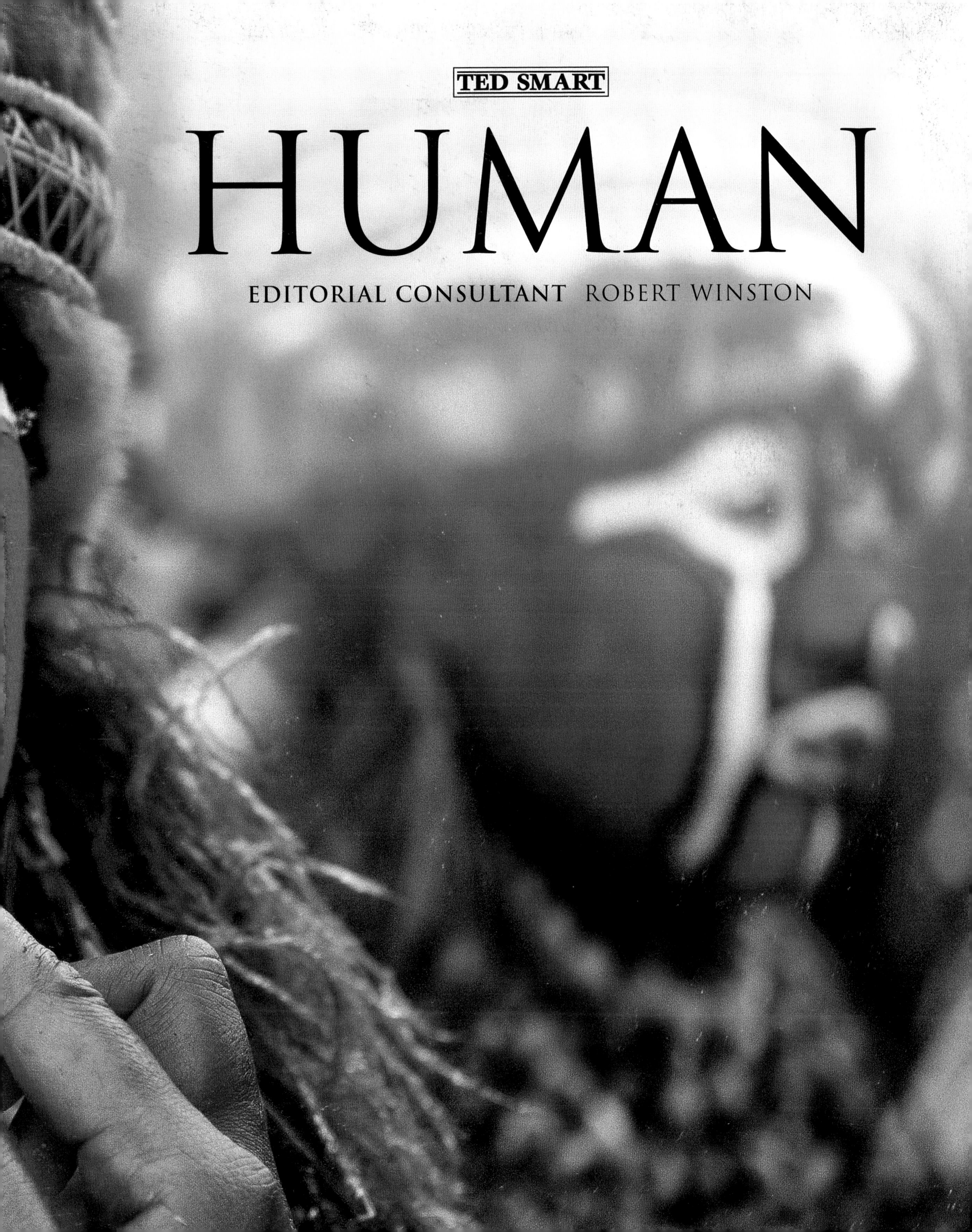

TED SMART

HUMAN

EDITORIAL CONSULTANT ROBERT WINSTON

LONDON, NEW YORK, MELBOURNE,
MUNICH AND DELHI

Senior Editor Janet Mohun
Project Editors Ann Baggaley, Joanna Benwell, Dawn Henderson, Rob Houston, Katie John
Indexer Hilary Bird

Editorial Manager Andrea Bagg
Publishing Director Jonathan Metcalf

Senior Designer Liz Sephton
Project Art Editors Sara Kimmins, Peter Laws, Maxine Lea, Mark Lloyd, Shahid Mahmood, Dan Newman
Design Assistants Iona Hoyle, Francis Wong
DTP Designer Julian Dams
Picture Researchers Gwen Campbell, Helen Stallion, Rob Nunn
DK Picture Library Romaine Werblow
Illustrators Joanna Cameron-Rutherford, Combustion Design and Advertising, Jane Fallows, Adam Howard, Kevin Jones Associates, Joern Kroeger, Debbie Maizels, Mirashade, Primal Pictures Ltd, 4site Visuals
Photographer John Davis
Cartographers Kenny Grant, Rob Stokes
Production Controllers Heather Hughes, Melanie Dowland

Managing Art Editor Marianne Markham
Art Director Bryn Walls

Smithsonian Project Co-ordinators
Ellen Nanney, Katie Mann

This edition produced for The Book People Limited
Hall Wood Avenue, Haydock, St Helens WA11 9UL

First published in Great Britain in 2004 by
Dorling Kindersley Limited
80 Strand, London WC2R ORL

A Penguin Company

A CIP catalogue record for this book is available from the British Library

ISBN 140530233X

Colour reproduction by Colourscan, Singapore
Printed and bound in China by Leo Paper Products

see our complete catalogue at
www.dk.com

CONTENTS

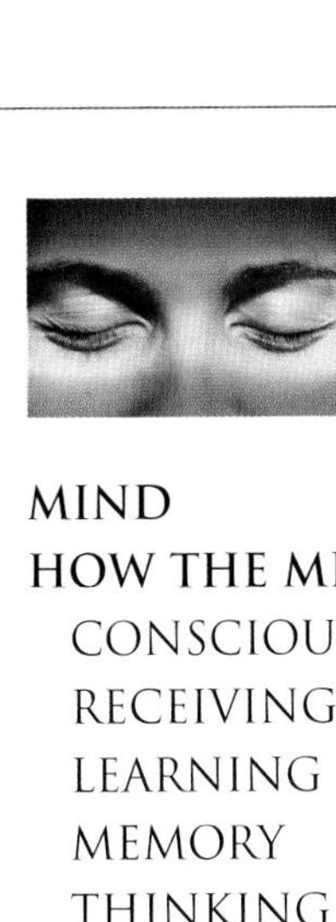

MIND

LIFE CYCLE

SOCIETY

CULTURE

PEOPLES

FUTURE

FOREWORD

This book is about what it means to be human. Humans are relative newcomers on the planet. Most animals have been on Earth for much longer than the 150,000 years or so that *Homo sapiens* has existed. Our ancestors were hominids, walking on two legs, coming from the open valleys of East Africa. We were an endangered species, and few in number. We were among the weakest of animals on the savanna – not particularly fleet of foot, no large teeth or claws as weapons, and with vulnerable babies needing constant nurture and protection.

But our species had, among other attributes, two remarkable advantages over other animals. One was our large brain, and with it, a keen intelligence and aptitude to adapt to hostile environments. The other was our sociability. Like many other apes we could not have survived without the ability to work in groups, and to cooperate in hunting and protecting each other. And human society was enhanced by our ability to communicate. Our hominid ancestors probably evolved from chimpanzeelike creatures between 5 and 15 million years ago. So it is not entirely surprising that modern humans are relatively similar to modern-day chimps. We share much of our DNA with them, and we probably have very similar genes. Yet the tiny genetic difference makes for a remarkable dissimilarity. We have a totally different physique, more powerful mental abilities, and an inbuilt skill at language.

Once the Earth's climate grew warmer and more stable around 10,000 years ago, humans were able to establish settlements. Human civilization went on to develop complex symbols and a written language. Now we have buildings and machines that mean we can live almost anywhere on Earth's surface – at high altitude, at extremes of temperature, even under water or in space. The story of evolution has largely been about how animals adapt to their environment by survival of the fittest. In one sense, we have become masters of our environment, and we now escape many evolutionary pressures.

Humans are phenomenally inquisitive – hence the development of science. We investigate and experiment, theorize about the origins of existence and the nature of the universe, and have a powerful spiritual sense. This aspect of the human mind, together with our aesthetic sense, led humans to develop pleasure in painting and music, poetry and theatre.

This book is one expression of humankind's natural inquisitiveness. No other species seems interested in how the body works. One great modern advance was the technology to make images of the brain at work, helping us understand the human mind – what makes each of us truly human.

We may all be from the same species, but humanity is very diverse. This book is testament to this amazing variety. Human development has proceeded in many ways in separate parts of the world. Each society found different ways to organize life – to survive, to educate, to dress, to govern itself, to communicate with the rest of the world, and to generate wealth. With rapid changes in technology, the internet, and increasing world trade and easy travel, we now live in a changing global society – where the consequences of our actions can have profound effects on people living at great distances.

So what of the future of the human species? It took 100,000 years to gain the ability to build crude dwellings, and even longer to learn to make stone tools. But in the last 200 years or so we have fashioned steam engines, harnessed this planet's stores of energy, ventured to the Moon, and learned the workings of the genome. Humankind's achievements are proceeding exponentially so it is impossible to predict our future. No doubt we shall improve our understanding of diseases and live longer. Maybe we shall even manipulate our own genes, changing the course of evolution. But fundamentally humans remain the fragile creatures they once were on the savanna. Even a modest change in Planet Earth could threaten our existence; and our use of technology, which has wrought good in the past, could see us lurch towards self-destruction.

It has been a privilege to be part of the team telling this remarkable story. I remain convinced that, while human existence is unlikely to continue untroubled, the nature of the human spirit and our moral sense means that we can be confident of human progress.

Robert Winston.

Robert Winston EDITORIAL CONSULTANT

ABOUT THIS BOOK

Human is divided into seven main sections plus a main introduction and a section on the future. Each of these seven sections is represented on the right. The first section, Origins, introduces human evolution and history. Next are sections on Body and Mind, which are followed by Life Cycle, the story of human life from birth to death. There are also sections on Society, Culture (including belief and language), and on Peoples around the world.

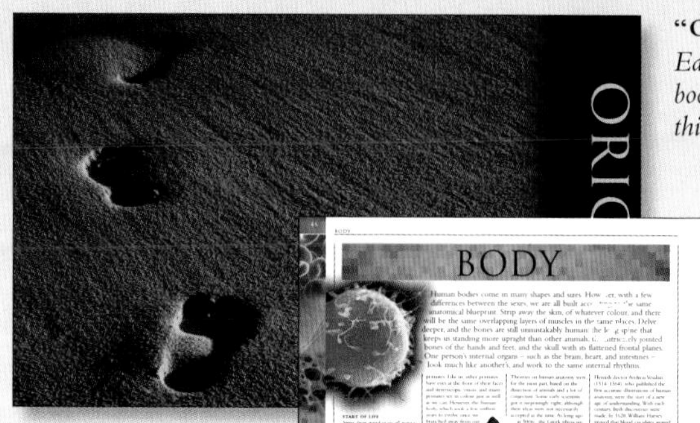

"ORIGINS" OPENER
Each of the eight main sections of the book is identified by a stunning image, this one from the Origins section.

BODY

"BODY" INTRODUCTION
Introductions to each section give a historical background, cover our current state of knowledge, and put the topic in the context of human life.

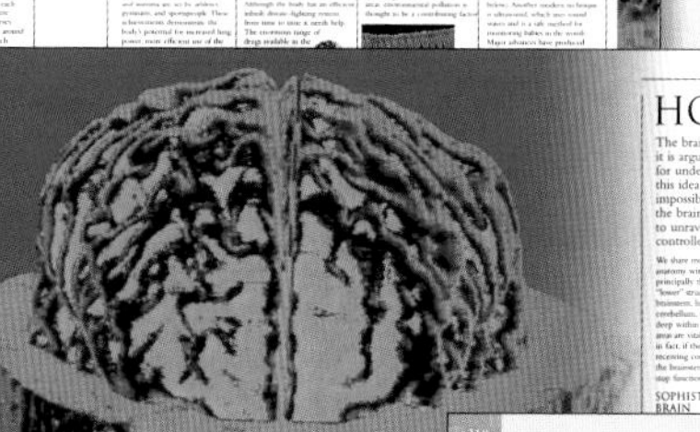

HOW THE MIND WORKS

"MIND" SUBSECTION INTRODUCTION
Subsection introductions have an eye-catching image on the lefthand page. The text gives an overview of the topics to follow. For example, this one for Mind introduces how we think, learn, and remember events.

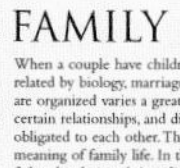

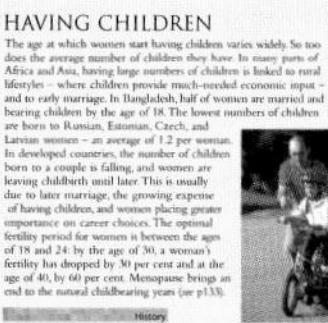

FAMILY

"LIFE CYCLE" EXPLANATORY PAGES
These pages on Family describe the general features of family life. They set the background to the pages that follow, which show individual types of family.

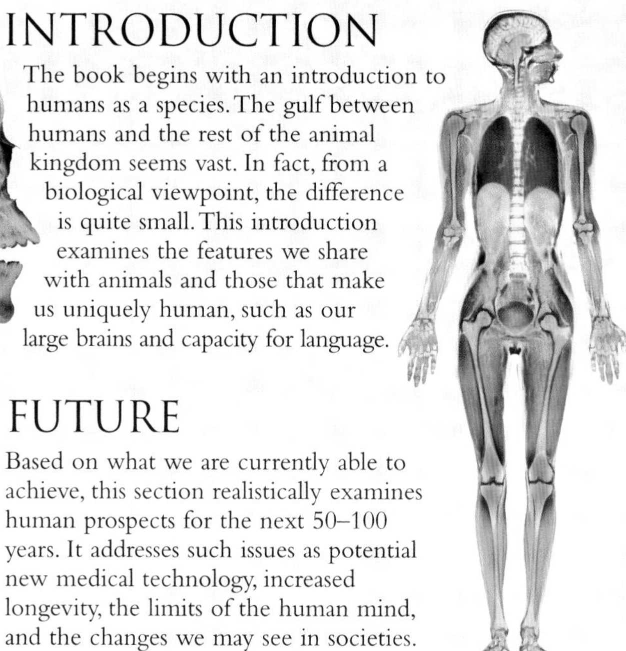

INTRODUCTION

The book begins with an introduction to humans as a species. The gulf between humans and the rest of the animal kingdom seems vast. In fact, from a biological viewpoint, the difference is quite small. This introduction examines the features we share with animals and those that make us uniquely human, such as our large brains and capacity for language.

FUTURE

Based on what we are currently able to achieve, this section realistically examines human prospects for the next 50–100 years. It addresses such issues as potential new medical technology, increased longevity, the limits of the human mind, and the changes we may see in societies.

Panel styles

Thematic panels

Throughout the book, five types of thematic panels appear, highlighting topics of special interest. These colour-coded panels are headed Fact, Health, History, Issue, or Profile. Fact panels include intriguing facts on subjects such as Tibetan "sky burials". Health panels describe disorders or aspects of lifestyle that can affect our health. History panels give a perspective on subjects such as ancient calendar systems. Profile panels include brief biographies of influential individuals such as Chinese poet Li Bo. Issue panels discuss some of the dilemmas we face in the 21st century.

Health
Positive health issues as well as some common disorders are covered in Health panels

Profile
Biographies of notable individuals give key dates and achievements

Fact
Interesting or unusual facts are picked out in Fact panels

History
These panels give the background to a topic with key dates

Issue
Topics of contention or debate are discussed in Issue panels

Health

A sedentary childhood

In the West the time that children spend watching television, surfing the internet, and playing video games is increasing. The average American child spends 1,023 hours a year watching television, and 6½ hours a day on all media-related activities. The amount of time spent on physical activity is falling. This is partly responsible for the rise in childhood obesity.

Profi

Sigmund Freud

Austrian psychoanalyst Sigmund Freu

Fact

The meaning of the veil

The origins of the wedding veil

History

Discovering germs

The discovery that certain types of microorganism cause diseases was

Issue

Where do we come from?

Some cultures believe in a single deity that created the world suddenly: many Christians believe God created Adam

Consultants and contributors

EDITORIAL CONSULTANT

Robert Winston Professor of Fertility Studies, Imperial College School of Medicine, London, UK

CONSULTANT FOR THE SMITHSONIAN INSTITUTION

Dr Don E Wilson National Museum of Natural History, Smithsonian Institution, Washington, DC, US

SECTION CONSULTANTS

Professor Frances Ashcroft Professor of Physiology, University of Oxford **Dr Sue Davidson** London **Dr Dylan Evans** Researcher and lecturer in Biomimetics, University of Bath **Dr Dena Freeman** Anthropologist, London **Professor Frank Furedi** Professor of Sociology, University of Kent at Canterbury **Professor Nick Humphrey** Psychologist and neuroscientist, Centre for Philosophy of Natural and Social Science, London School of Economics **Professor Tom Kirkwood** Professor of Gerontology, Institute for Aging and Health, Newcastle University **Dr Graham Ogg** Institute of Molecular Medicine, Oxford **Dr Max Steuer** Economist, London School of Economics **Professor Christopher Stringer** Department of Palaeontology, Natural History Museum, London

CONTRIBUTORS

Susan Aldridge, Jo de Berry, Susan Blackmore, Rita Carter, David Brake, Liam D'Arcy Brown, Adam Burgess, Andrew Chapman, John Coggins, Carol Cooper, Ludovic Coupaye, Andrew Dalby, Robert Dinwiddie, Roger East, Dorothy Einon, Dena Freeman, Peggy Froerer, Ken Gilhooley, Amra Hewitt, Ben Hoare, Alan Hudson, Mark Jamieson, Dawn Marley, Magnus Marsden, Wim Mellaerts, Rory Miller, Zoran Milutinovic, Ben Morgan, Phil Mullan, Brendan O'Neill, Katie Parsons, Mukul Patel, Dr Penny Preston, Hazel Richardson, Sanna Rimpilainen, Nigel Ritchie, D J Sagar, Martin Shevrington, John Skoyles, Max Steuer, Richard J Thomas, Edmund Waite, Helen Watson, Philip Wilkinson, James Woudhuysen, Eve Zucker, Cambridge International Reference on Current Affairs (CIRCA)

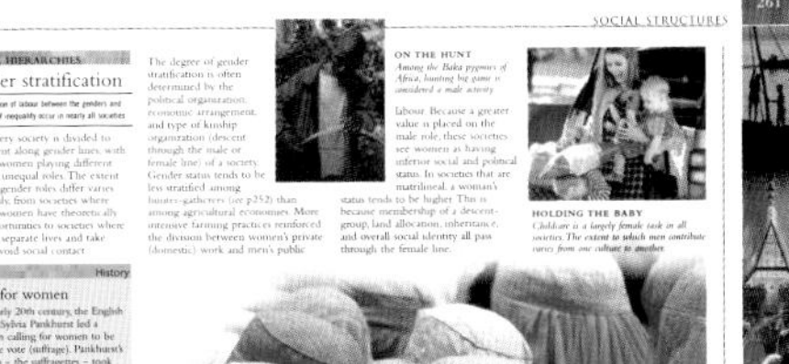

"SOCIETY" PROFILE PAGES

Profile entries provide information on individual aspects of the subject covered. For example, within the Society section, types of social hierarchy, power structure, and conflict are covered.

MAP AND FACT FILE

Each entry in the Peoples section is headed by a map, pinpointing the main areas of distribution, along with details of population, languages spoken, and main belief systems.

Map

Location of peoples is pinpointed on specially prepared relief maps

Location

Profiles are arranged on the page by geographic location

Fact file

Each profile is headed by a fact file listing location, population, languages spoken, and beliefs

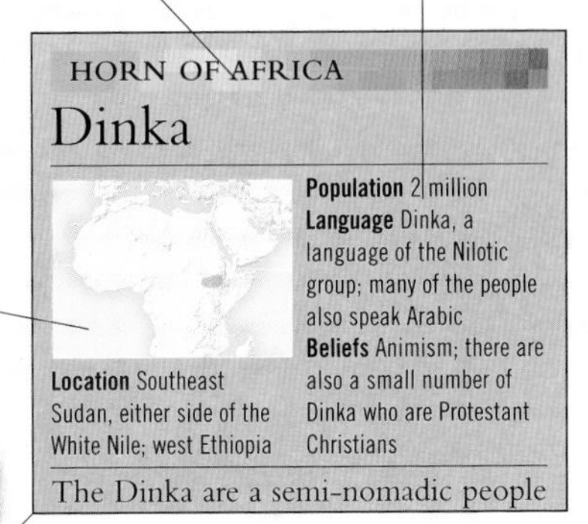

HORN OF AFRICA
Dinka
Population 2 million
Language Dinka, a language of the Nilotic group; many of the people also speak Arabic
Beliefs Animism; there are also a small number of Dinka who are Protestant Christians
Location Southeast Sudan, either side of the White Nile; west Ethiopia
The Dinka are a semi-nomadic people

"PEOPLES" PROFILE PAGES

More than 250 profiles of peoples from every part of the world are presented. Each profile begins with a location map and fact file.

"CULTURE" PROFILE PAGES

This section looks at cultural variations, including religions, clothing, and languages. The illustrated profiles here are from pages on types of clothing and adornment. Most profiles are illustrated with photographs.

INTRODUCTION

INTRODUCTION

One of the curious things that makes our species different from others is that we can recognize ourselves in a mirror. To scientists and philosophers, our capacity to understand a reflection is a sign of one of our most important distinguishing features: self-awareness. Only the most intelligent animals, including chimps and gorillas, show hints of this very peculiar ability. Self-awareness not only defines us, it also drives our ongoing efforts to understand our very nature. Since the beginning of history, people have struggled to unravel the mystery of human nature and find out exactly what makes us so special.

HUMAN BRAIN
The human brain is about three times bigger than it should be for an average primate of our body size. Our large brain underpins many of our unusual mental abilities.

Humanity has not always seen itself as part of the animal kingdom. For centuries, humans thought of themselves as higher beings with souls, free will, and consciousness, and saw animals as mindless creatures driven by instinct. As recently as the 19th century, Charles Darwin caused uproar by saying we were descended from apes, and even today the word "animal" retains its historic meaning: something base, violent, and inhuman.

For biologists, however, "human" and "animal" are far from opposites. Our species, *Homo sapiens*, is unequivocally an ape. Peel away the distracting layer of clothes and the peculiar, naked skin, and we have exactly the same complement of organs and tissues as our ape cousins. To a geneticist, the difference is even smaller: human and chimp DNA differs by only 1–2 per cent. So the evidence from science says that humans are unquestionably animals, but common sense and tradition tell us there is a gulf between us and the rest of the animal world. This apparent contradiction lies at the heart of the mystery of human nature, and only by resolving it can we begin to understand what it means to be human. To do that, we need to understand our place in the animal kingdom.

THE ANIMAL KINGDOM

We are just one of at least 1.5 million species that make up the kingdom Animalia, one of the five kingdoms of life (*see* The tree of life, below). In the past, zoologists arranged all these animal species into a family tree with two major branches: vertebrates (all the animals with backbones, including us) and invertebrates (worms, insects, spiders, and so on). Over the years, as more information was amassed about the evolutionary history of the animals, our branch of the tree shrank in importance and the classification changed. Today, vertebrates make up a mere twig on one of 30 or so main branches, or "phyla", in the animal kingdom. Despite this apparent relegation, vertebrates are the dominant animals. Squeezed onto that small twig in the tree of life are all the world's mammals, birds, reptiles, amphibians, and fish, and among their number are the biggest and most spectacular creatures ever to have lived.

A TYPICAL VERTEBRATE

The first vertebrates, our distant ancestors of 400 million years ago, were fish. The legacy of those aquatic ancestors is still with us, including the defining vertebrate feature: a bony spine, flanked by pairs of muscles. This feature gave the first fish superb mobility in water, making them the oceans' top predators. Today, it

MAN ON THE MOON
The first moon landing in 1969 is just one example of the human drive to explore and understand the universe. A total of 12 astronauts walked on the Moon during the six Apollo missions.

The tree of life

All living things can be classified into a tree of life that reflects their evolutionary history. At the base of the tree are the five kingdoms of life, each of which can be subdivided into a branching network of many different species. This chart shows in a very simplified form how our own species fits into the tree. Working from left to right, the species become increasingly closely related to each other. Like all species, humans have a scientific name made up of two parts: a genus and a species. We are the sole surviving members of the genus *Homo*, which also includes our direct ancestor *Homo erectus* and close cousin *Homo neanderthalensis*. Humans are also the only surviving members of a subfamily of apes now known as hominins (family Homininae), made up of species of ape that lived on the ground and moved around on two feet rather than "knuckle-walking", as gorillas and chimpanzees do. Hominins are believed to have split from the chimpanzee branch of the ape family tree about 5–6 million years ago.

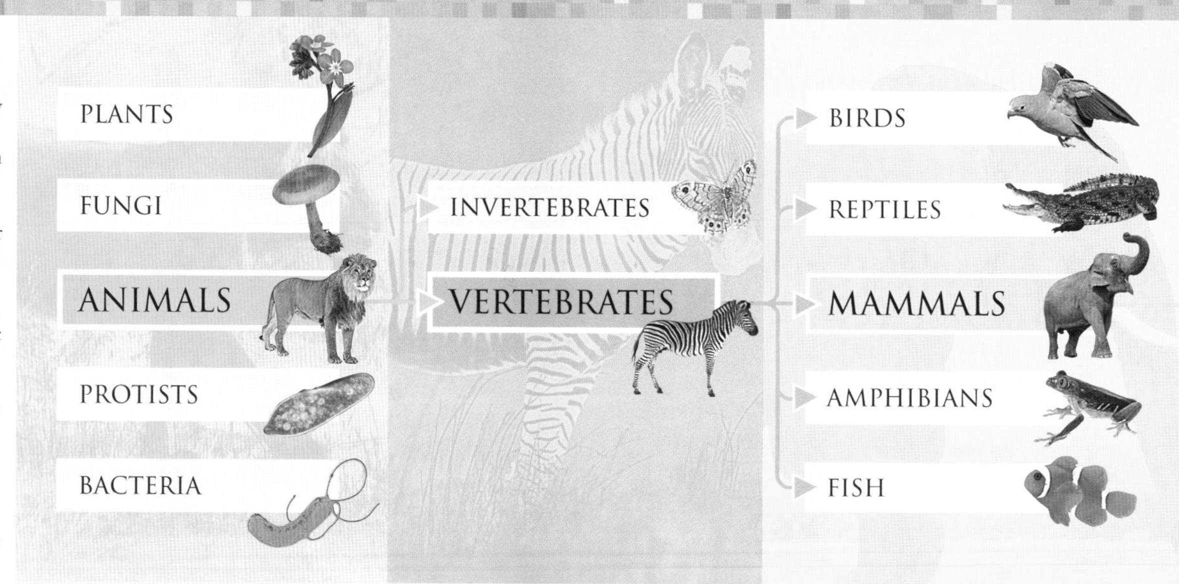

KINGDOMS OF LIFE
Living things can be divided into five main kingdoms. Our species belongs to the animal kingdom, along with about 1.5 million other known species.

ANIMAL DIVISIONS
Animals have traditionally been divided into those with a backbone (vertebrates) and those without (invertebrates). All vertebrates have a similar skeletal plan.

VERTEBRATE CLASSES
Vertebrates are divided into five classes. We belong to the mammals, which are defined by the ability to produce milk. Most mammals also have hair or fur.

provides the main supporting strut in the human skeleton, allowing us to stand and move. Beneath our skin, our body's tissues are still bathed in a salty solution, the chemical composition of which is surprisingly similar to that of seawater. This is yet another reminder of an aquatic past.

Except for our unusual upright stance, human anatomy is typical for a land vertebrate. We have the same skeletal arrangement as most other land-living vertebrates, with four jointed limbs that each end in a spread of five digits, the typical number for vertebrates. Our digits form our hands and feet, but those of other vertebrates have taken many different forms. A bat's digits, for example, form the bony struts of its wings, and a horse's single digit (all the others have withered away) ends in a grotesquely enlarged toenail – its hoof.

Humans have the typical sense organs of a vertebrate, including a single pair of eyes and a single pair of ears. The stream of data that comes from these organs is processed by another feature that is typical of vertebrates: a brain enclosed in a strong, bony case.

The tail that is characteristic of the majority of vertebrates appears and disappears before humans are born, and the gill slits present in aquatic vertebrates appear only fleetingly when we exist as embryos – one has evolved into the eustachian tube, a narrow airway that connects the middle ear to the throat.

HUMANS AS MAMMALS

Further along the tree of life, the vertebrate twig splits into a spread of smaller twigs, which are known as classes. We belong to the class Mammalia – the mammals. There are several defining features that set mammals apart from other vertebrates, the most important of which is that mammals produce milk to nourish their young (the word "mammal" comes from the Latin *mamma*, which means "breast"). Like other mammals, we are warm-blooded, and our bodies are covered by hair. Our species belongs to a category of mammals known as primates. Nearly all primates are tree-dwellers, and most are confined to the belt of warm forest that girdles our planet between the tropics. Many aspects of human anatomy – for example, our grasping hands, forward-facing eyes, and colour vision – are typical primate features, vestiges of millions of generations spent clambering through trees. And we are unmistakably primates in our social lives and behaviour: like many monkeys and apes, we live in complex, hierarchical societies where survival depends on navigating an ever-changing web of social relationships.

Fact

Ninety-nine per cent chimp?

Scientists have used a plethora of techniques to compare human genes to those of chimpanzees and other apes. The latest experiments suggest we differ from chimps in just 1.2 per cent of our active genes. Indeed, some scientists say humans, chimps, and the closely related bonobo should all be placed in the same genus, *Homo*. However, although the genetic distance between humans and apes is tiny, the differences in appearance, behaviour, and intelligence are profound. The few genes that make humans different probably include key controller genes that affect the activity of many other genes during our development, with far-reaching consequences for our bodies and brains.

THE DNA CONNECTION
Deciphering the genetic code of living creatures enables scientists to understand the links between species. Shared DNA suggests shared ancestors.

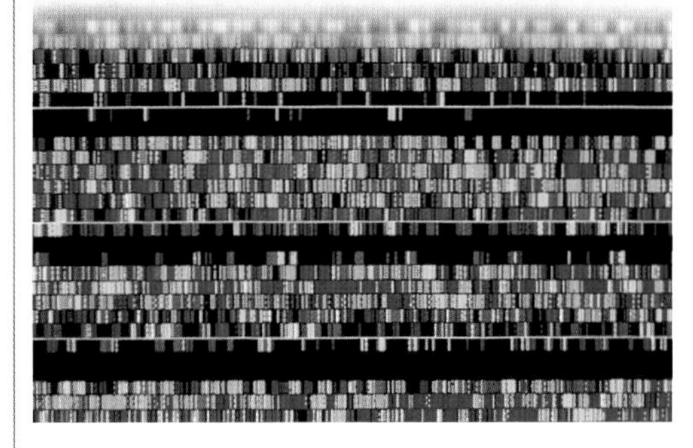

BEING HUMAN

It is common knowledge that we share nearly all our genes with chimpanzees. Yet the tiny fraction that makes us different has a profound effect on our anatomy, so much so that, for centuries, scientists had difficulty believing we had evolved from apes at all. Unlike other apes, we lack the opposable (flexible) big toes needed for climbing, and we get about by walking on outsized hind limbs with our short forelimbs dangling in the air. Compared to apes, our skin is almost naked, our skeleton is bent out of shape, our head is swollen like a balloon, and our breasts and penis are bizarrely large. Even so, if alien scientists were to land on Earth and study *Homo sapiens*, they would have

MAMMALIAN ORDERS
There are around 21 "orders" of mammals. We belong to the primates, most of which are tree-dwellers with grasping hands and large brains.

PRIMATE CATEGORIES
Primates are divided into three main groups: apes, monkeys, and prosimians. Apes and monkeys are active by day but many prosimians are nocturnal.

THE APES
Apes are characterized by their large, muscular arms, their mobile shoulders, and the lack of a tail. All but humans are restricted to areas of tropical forest.

very little difficulty in recognizing us as apes. Physically unusual we may be, but the mountain of similarities between us and the other apes would soon become apparent.

If those visitors were to study the human mind, however, they might think that we too had arrived from another planet. In intellectual terms, a yawning chasm separates *Homo sapiens* from every other species on Earth. We have language of astonishing complexity. We build cities, cars, spaceships. We invented morality, religion, trade, science, world wars. We can think symbolically, create art, plan for the future, and solve problems using our imagination. And, perhaps even better than other primates, we can read each other's minds and intentions from only the slightest inflection of the voice or dart of the eyes.

FLEXIBLE SHOULDERS

Many of the features that make humans special are unique to us, but others are shared with our close relatives. Like other apes, we have flatter chests and longer, more powerful arms than monkeys. These differences evolved because of the way that apes move through the trees. Instead of scurrying along branches on all fours as monkeys do, apes haul themselves up trees by grasping the trunk in their arms. The shoulder blades of apes are at the back of the chest rather than the sides, which frees up the shoulder joint and allows the arm to reach overhead and swing around. In humans, the flexible shoulders that evolved for climbing are put to other uses. Uniquely, we can throw objects with force and accuracy – an ability that proved enormously useful when our hunting ancestors invented throwable weapons. Flexible shoulders also make our grasping hands more useful. If our shoulders were less mobile, we would find it much harder to carry and manipulate objects.

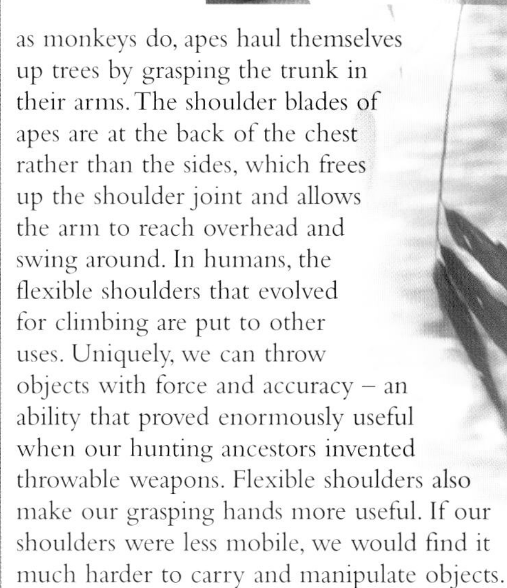

SHOULDER ACTION
Humans and other apes have highly flexible shoulders that give us free overarm movement. As a result, humans can throw a javelin with accuracy, and orangutans are able to swing between branches, hanging from their hands.

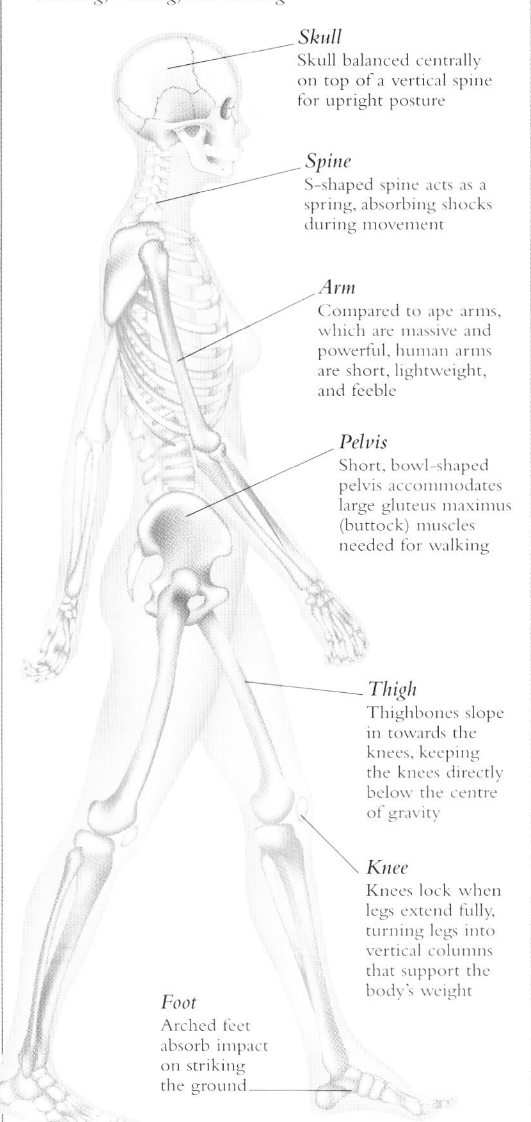
Skull
Skull balanced centrally on top of a vertical spine for upright posture
Spine
S-shaped spine acts as a spring, absorbing shocks during movement
Arm
Compared to ape arms, which are massive and powerful, human arms are short, lightweight, and feeble
Pelvis
Short, bowl-shaped pelvis accommodates large gluteus maximus (buttock) muscles needed for walking
Thigh
Thighbones slope in towards the knees, keeping the knees directly below the centre of gravity
Knee
Knees lock when legs extend fully, turning legs into vertical columns that support the body's weight
Foot
Arched feet absorb impact on striking the ground

BUILT FOR WALKING
Evolution has dramatically re-engineered the human skeleton since we split from the rest of the apes, changing a body plan adapted for climbing into one adapted for standing, walking, and running.

ON TWO FEET

The ability to walk on two feet not only sets humans apart from other apes but marks a major divide between us and the rest of the mammals. No other mammal can match our ability to stand, walk, and run on two legs. Of course, many other animals are capable of two-legged – or "bipedal" – movement, including ostriches, kangaroos (although they hop), and penguins. However, unlike ostriches and kangaroos, which are counterweighted by long necks and tails that act like a tightrope walker's pole, humans stay vertical largely because our nervous system is so beautifully coordinated. Our sense of balance also enables us to learn how to ice-skate, ski, and even perform feats such as walking on our hands. One price we pay for moving in such a strange manner is that walking takes time for us to master – human babies cannot walk for about a year.

Although other apes cannot walk or run as we can, they show hints of our ability. Apes have a more upright posture than monkeys. In trees, they frequently stand on their hind limbs while holding branches with their arms. Chimps and gibbons can even walk – or waddle – a short distance on the ground on two feet, an ability that comes in handy when they need to wade across rivers, but one that requires great effort.

Bipedal motion is easy for humans as a result of changes in the skeleton of our ancestors. We can extend our legs fully to form a vertical column that supports the body's weight, and the knees lock to prevent overextension of the lower leg. In contrast, apes can only partially extend the lower leg, forcing them to stand with their knees bent and rely on muscle power alone to stay upright, which is tiring.

Seen from the front, the human femur (thighbone) slopes inward from the hip to the knee, ensuring that the knees and feet are directly below the body's centre of gravity. Apes'

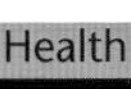
Health

An evolutionary compromise

During childbirth, a baby must pass through the space in the middle of the mother's pelvis. In most mammals this is a straightforward process, but in humans it is exceptionally painful, dangerous, and slow. Because we evolved a two-legged posture, we have a much narrower pelvis and a smaller pelvic outlet than other apes. However, human babies also have an unusually large head to accommodate the larger human brain. As a result, humans are born at a relatively early stage of their development, at which time they are physically helpless and therefore entirely dependent on their parents. Another evolutionary compromise is that women need a wider pelvis than men, which makes them slower runners, and slightly less athletic.

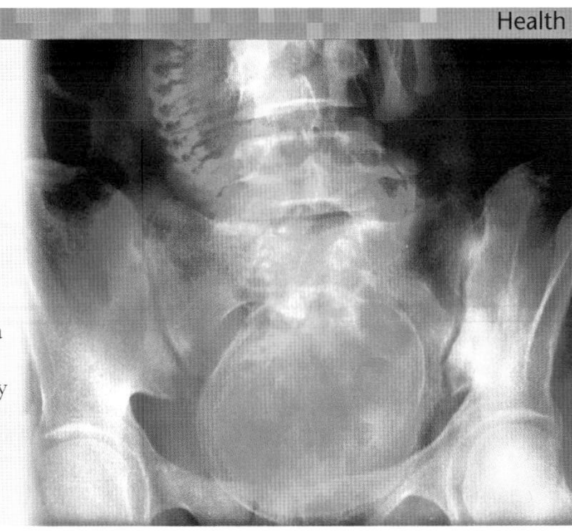

X-RAY OF CHILDBIRTH

legs are more splayed, resulting in an awkward, waddling gait. Human legs are much larger than our arms, giving our body a relatively low centre of gravity that aids stability. Our centre of gravity falls between the two hips, giving us a stable, vertical posture. In contrast, apes have large, muscular arms, short legs, and a high centre of gravity in front of its hips. An ape's high centre of gravity results in an unstable, crooked posture when upright.

Our feet are arched so that the heel and the ball of the foot carry our weight as we move. In contrast, apes stand with the whole length of the foot on the ground. Apes have opposable big toes for climbing, but in humans the big toe is aligned with the others. When we walk, weight is transmitted from the heel to the ball of the foot and on to the big toe, which is the last point of contact as the foot pushes off the ground. This is a more efficient way of walking.

The human skull is balanced on top of a vertical spine rather than being held by muscles in front of a horizontal spine. As a result, the hole through which our spinal cord passes (the foramen magnum) is shifted forwards relative to that of apes, so that it lies directly under the brain. The spine is curved into an undulating S-shape that acts as a spring, absorbing shocks during movement.

To accommodate the enormous muscles that we need for walking – especially the gluteus maximus (buttock) – the human pelvis is much shorter and broader than that of other apes. It is also bowl-shaped to support the abdominal organs cradled above it.

There are drawbacks to being bipedal. One is that childbirth is more painful and protracted in humans than in other mammals (*see* An evolutionary compromise, p14). Another is that lower back pain and injury are more common in humans. The vertebrae of the lower back have to carry all the weight of the upper body, yet they must be small to preserve the spine's flexibility. When we bend over to lift a heavy weight, the lower back is subjected to a force much greater than our body weight, and this force can rupture one of the shock-absorbing discs between the bones (called a slipped disc).

HANDS FREE

One of the great advantages of being upright is that it frees the hands. Like other primates, we have opposable thumbs – digits that move in the opposite direction to the rest. This ability turns hands into pincers that can grasp and manipulate objects. Gorillas use their hands to pick apart thorny plants and uproot edible material, and chimps use theirs to handle simple tools, such as twigs with which to fish termites or rocks to crack nuts. Unlike humans, though, these apes also have to use their hands as feet, which limits how delicate and nimble the fingers can become. Human hands, being completely free, have evolved into precision instruments of amazing dexterity. Our hands can tie shoelaces, play pianos, thread needles, grasp hammers, and count coins in a pocket.

Human fingertips are packed with receptors sensitive to pressure, and they are backed by nails instead of claws, improving the sense of touch. The palms of our hands are among the only parts of our skin that are truly hairless (along with the soles of our feet and our lips), which improves our grip. To improve grip still further, the skin of our fingertips is roughened by tiny ridges bearing oil and sweat glands that keep the skin supple and damp, allowing us to pick up and grip the tiniest objects. At the same time, the part of the brain controlling hand–eye coordination is far more highly developed than in other primates. Manual dexterity is crucial to our species. Without it, we would be unable to make fire, throw spears, build houses, or invent endless varieties of tools.

Fact

Slow but economical

Walking on two legs may have evolved to save us energy – we use less energy walking on two legs than do chimps or gorillas when they "knuckle-walk" on all fours. However, compared to most four-legged mammals, we are very poor runners. We burn through calories at an extravagant rate while running, and our top speed is unimpressive. Over a distance of about a mile, the best athletes average about 15mph (24km/h). Horses, dogs, and antelope can easily exceed 30mph (48km/h) and can sustain their speed for much longer.

SCENTS AND SMELLS

Most mammals live in a world of smells. They use scent glands to annoint landmarks in their territory, leaving long-lasting signals that attract mates or repel rivals. Scents contain a wealth of information, telling visitors the age, sex, identity, and reproductive status of the marker.

For primates – and especially for us – the sense of smell is less important. Our olfactory bulbs (the parts of the brain that process smell) are a fraction of the normal mammalian size. Even so, we do produce scents that have social and sexual significance. Our scents are made in the hairiest parts of the adult body: the armpits and groin. Special sweat glands in these areas exude a viscous, strong-smelling type of sweat called apocrine sweat that contains chemicals known as pheromones. The composition of apocrine sweat varies according to the menstrual cycle and mood. Before the era of deodorants and perfume, apocrine sweat probably helped people choose their sexual partners. Scientists think that the complex odours contained in the sweat somehow convey information about whether a potential partner's immune system will be compatible with our own.

VISION

Although smell and hearing are the primary senses for most mammals, vision is the dominant sense in humans. As primates, we have excellent colour vision and the ability to perceive tiny details, owing to the way light-detecting cells

A SUPERB SENSE OF BALANCE

Our skeletal structure and sophisticated nervous system enable us to keep our balance even when carrying objects that are bigger than ourselves.

are packed into the back of the human eye. In the centre of the retina (the light-sensitive membrane at the back of the eye) is a small pit called the fovea, where light-sensitive cells are crammed together very densely to create the detailed central point of the visual field. Only animals with a fovea have very sharp vision; these include birds of prey and primates. The human fovea is packed with colour-detecting cells called cone cells, of which there are three types: one for each primary colour, giving us full colour vision. Most mammals have only one or two types of cone cell, making them colour blind in the medical sense.

Excellent colour vision comes at a price: cone cells work only in daylight, and we are almost blind at night. Poor night vision is one of the reasons that humans and other apes spend the night asleep indoors or in treetops, far from nocturnal predators that can see in the dark.

Like all primates, we can see in 3-D. Our primate ancestors evolved three-dimensional vision partly because it helped them move in the treetops. For us, 3-D vision is more important for manipulating objects. This type of vision requires overlapping fields of view, with both eyes facing in roughly the same direction (quite unlike the sideways-facing eyes of animals such as rabbits, which need all-round vision to spot approaching predators). Humans are typical primates in this respect. Our eyes face forward, giving us a wide area of 3-D vision, and our nose is short to minimize obstruction of the field of view. To compensate for our lack of

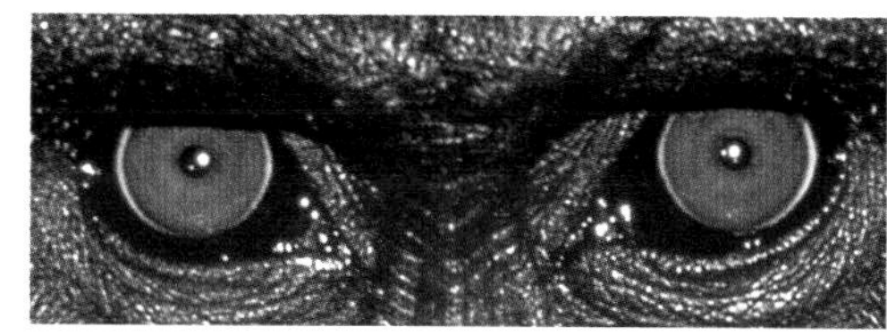

GORILLA

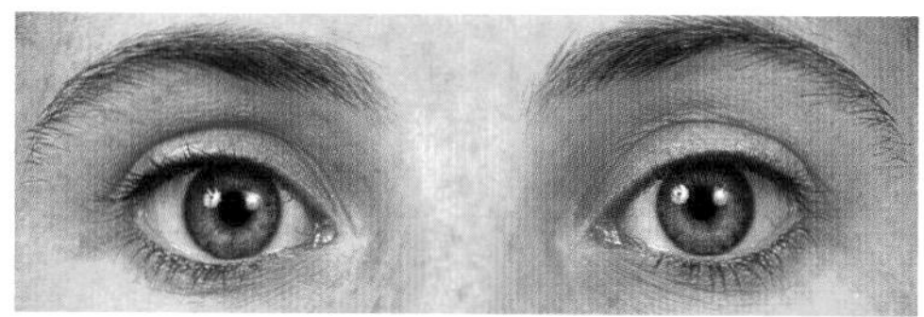

HUMAN

WHITE EYES

Compared to apes, human eyes have a larger white zone, making direction of gaze obvious to observers and so aiding communication involving eye contact.

all-round vision, we have mobile eyeballs and necks, allowing us to look around without turning the body. In one respect, our eyes are different from those of other primates: we have a large white zone around the iris that makes our eyes conspicuous and allows us to communicate direction of gaze. Most apes have largely brown eyes that camouflage their gaze.

KEEPING WARM

Like all mammals, humans are "warm-blooded", which means that our bodies can maintain a consistently warm internal temperature. Reptiles and amphibians, in contrast, are "cold-blooded", which is a somewhat misleading term since these animals sometimes have warmer blood than mammals do. The crucial difference is that cold-blooded animals cannot maintain a constant body temperature independent of the conditions around them; they are therefore at the mercy of their environment. Desert lizards, for example, become sluggish at night as the temperature plummets. In the mornings, they need to sun themselves for some time in order to become active again. Mammals, however, can stay active even in Arctic winters. Of all the mammals, humans have made the most of the adaptability that warm-bloodedness brings.

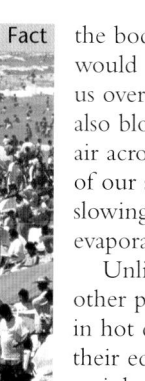

NEARLY NAKED

For most mammals, hair is a vital part of the warm-blooded heat-retention system. A dense layer of hairs (fur) traps warm air next to the body like a blanket, slowing down the loss of precious body heat. Only in the largest mammals, such as whales and elephants, does hair become redundant, because their huge bulk and resulting low surface area slows down the loss of heat. Although human skin appears hairless, we are not truly naked – we have just as many hairs as chimpanzees. The difference is that our hairs are much smaller and finer. With the exception of the patch of thick hair on the head, our body hair is of little use as insulation.

Why, then, are humans almost naked? The answer probably lies in our historical need to stay cool. When our ancestors left the forests for more open savannas, they were exposed to the full force of the stifling African sun. Only the most heat-tolerant animals are able to stay active during the hours of daylight in the African savanna. We could tolerate such conditions thanks to our upright stance (which minimizes exposure to the sun), our bare skin, and a sweat system more advanced than that of any other mammal.

Our skin is covered with millions of glands that produce a watery secretion called eccrine sweat when our body temperature rises. This sweat evaporates quickly from the skin, drawing heat away from the body. A layer of fur would not only make us overheat but would also block the flow of air across the surface of our skin, thereby slowing down the evaporation of sweat.

Unlike humans, other primates stay dry in hot conditions, and their eccrine glands are mainly restricted to the bare skin on their palms and soles, where the skin needs to be kept moist to improve grip and sensitivity. Our palms also have high numbers of eccrine glands that, just as in other primates, become active in times of stress – a vestige of our tree-dwelling past when good grip was vital for making a quick escape.

When our ancestors left Africa and spread to other parts of the world, they remained tropical in their biology and had to adapt to the cold by using their ingenuity. With help from clothes, fire, and shelter, they conquered almost every environment on the planet. Evolution did not stop altogether, however. In bitterly cold parts of the world, such as the Arctic, people seem to have evolved a shorter, more compact build that helps retain heat. People native to much hotter, drier climates, such as the Masai and the Arabs, typically have a more slender, long-limbed build that increases the body's surface area and so helps shed heat.

TEETH AND DIET

To fuel their on-board central heating systems, mammals have to consume around 10 times more calories than do cold-blooded animals, and humans are no exception. Much like other mammals, we have a highly efficient digestive system and sophisticated teeth that mesh together precisely to grind up food like

INUIT

MASAI

CLIMATE CONTROL

The Masai have a slender build that sheds heat; the Inuit have a compact frame that retains heat.

Fact

Our love of water

Our desire to be near water may have evolutionary roots. Early humans lived on the African savanna, where they came to tolerate the heat by evolving bare skin and producing huge amounts of sweat. This allowed them to be active during the hottest hours of the day, when lions and other predators usually rest. However, our ancestors had to stay near water to replenish the fluid lost through sweat. Until they invented flasks made from animal skins and ostrich eggs, early humans probably never strayed far from waterholes or rivers. Water sources were also a magnet when people first began building settlements.

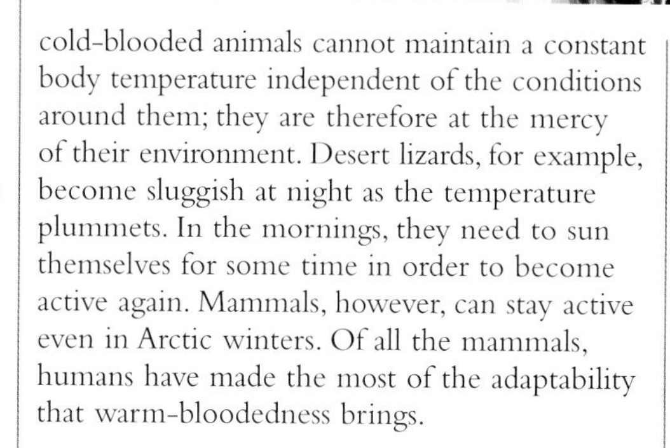

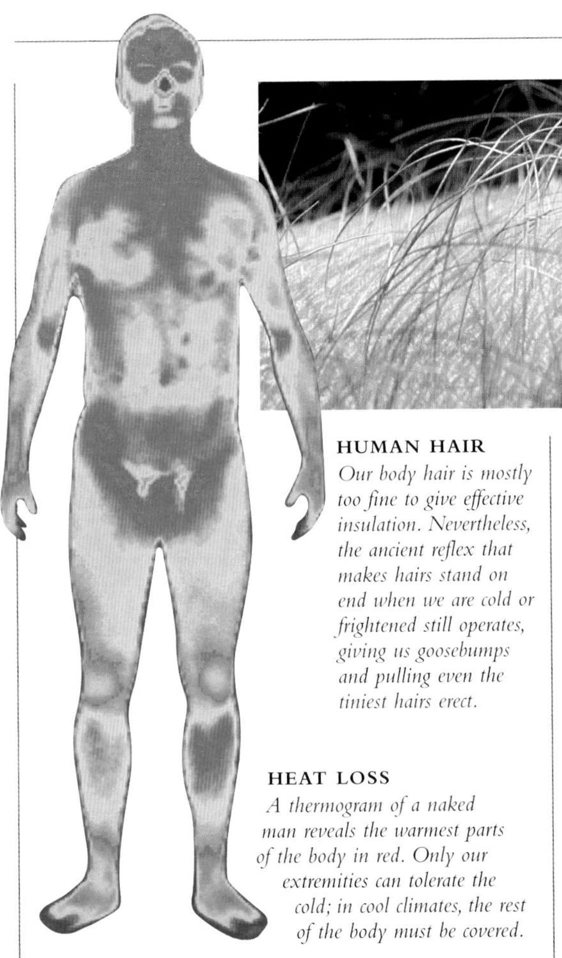

HUMAN HAIR
Our body hair is mostly too fine to give effective insulation. Nevertheless, the ancient reflex that makes hairs stand on end when we are cold or frightened still operates, giving us goosebumps and pulling even the tiniest hairs erect.

HEAT LOSS
A thermogram of a naked man reveals the warmest parts of the body in red. Only our extremities can tolerate the cold; in cool climates, the rest of the body must be covered.

the stones of a flourmill. Whereas reptiles shed and regrow their crude, peglike teeth throughout life, mammals typically grow just two sets of more finely-tuned teeth: a set of deciduous (milk) teeth and an adult set.

Some primates specialize in eating leaves, insects, or fruit; others are generalists, able to exploit a diverse range of foods, including fruit, leaves, seeds, insects, and small animals. Generalists are among the most inquisitive and intelligent of the primates. They are able to adapt to new situations, discover new types of food, and use their cunning and ingenuity to overcome the biological defences of the animals and plants they eat. Humans belong to this generalist category. Our teeth are small and relatively unspecialized, indicating that our natural diet includes a diverse range of foods rather than a single staple. We have the same dental formula as other apes: 8 incisors, 4 canines, 8 premolars, and 12 molars. However, our teeth are smaller because we can use our hands, tools, or cooking to deal with tough plants and meat. Some experts think that early humans were scavengers rather than hunters, but most agree that animal foods have long been an important part of our diet. Today, the amount of meat in the human diet depends mainly on where people live. In tropical countries, where plant foods are available all year round, starchy foods like grain and plant roots make up the bulk of the diet, supplemented by small amounts of meat. Further north, where the seasonal climate makes plant foods hard to find in winter, meat and milk are more important. And some Arctic people survive on almost nothing but meat.

SEX LIVES

Ape species differ enormously in their sex lives. In gorillas, a dominant male lives with a harem of female mates, but in chimpanzees, one female may mate with every single member of a male coalition. Gibbons are monogamous; orangutans are solitary; and bonobos have casual sex as a greeting. Humans follow none of these patterns. Biologists describe the human mating system as "mildly polygynous", which means that we are nearly monogamous but sometimes have a tendency towards promiscuity.

Compared to other apes, our sexual anatomy is unusual. The human penis – the longest and thickest of any primate's – is about four times larger than a gorilla's when erect, yet human testes are smaller than those of chimpanzees. Women's breasts are not only larger than those of other female apes, but are even larger than is necessary to produce milk. Why these differences exist is not entirely clear, but many scientists think they are caused by "sexual selection", an evolutionary process in which animals (usually males) acquire sexual adornments, such as the peacock's tail or the stag's antlers.

Like all higher primates, we have sex for pleasure, even when women are not ready to conceive. Chimpanzees and many other primates have an area of bare genital tissue that swells and colours during ovulation, advertising a female's fertility. In our species, however, a woman's readiness to conceive is hidden even from herself. Some biologists think that this "concealed ovulation" evolved in women to make men more faithful – if ovulation was signposted, a man would safely be able to neglect his partner during her infertile period in the knowledge that any offspring would be his. According to this theory, concealed ovulation and sex for pleasure are among a suite of adaptations that cement the bond between couples and encourage fathers to share in the childrearing process. However, another possible explanation for concealed ovulation is our upright posture. When humans became upright, the external female genitals changed position and became concealed between the legs. Upright posture may also explain the human male's long penis.

LIFE COURSE

As mammals, we are typical in looking after our offspring and feeding them milk (many other animals simply abandon their eggs or offspring to fate), but we care for our offspring for much longer than other mammals do. Humans are not only born helpless but develop slowly and stay dependent on parents for an extraordinarily long period. Gorillas and chimps are able to feed themselves as soon as they are weaned, but human infants continue to be fed for years after weaning. Our period of dependency is longer because our brains need more time to grow, develop, and learn the complex social and technical skills needed for survival. Because young children need so much care, we usually live in monogamous families, with both parents sharing the burden of providing food.

As well as growing up slowly, humans live longer than other apes. Our life expectancy is about 80 years, compared to 40–50 years in chimps. Human females lose their fertility at around age 50, yet survive for another 3 decades or so – much longer than any other mammals survive after the end of the fertile period. The menopause seems to play an important role in our species' life course and family structure. By looking after their grown-up daughters and their grandchildren, older women may contribute more to their families' success than would be the case if they continued to bear

THE ROLE OF GRANDMOTHERS
Grandmothers play a crucial role in the human life cycle by helping their offspring to raise families. The importance of grandmothering may have led to the evolution of the long human lifespan and early menopause.

Issue

Is it natural to eat meat?

There is great debate about whether it is "natural" for humans to eat meat. Some people say that since meat-eating is rare in primates, the amount of meat in the human diet is unnatural. Moreover, we lack the large canine teeth and claws of natural carnivores. However, chimpanzees, our closest relatives, eat meat, and our ancestors probably ate even more meat than we do when they moved from the forests to game-rich savannas. Our teeth may be small because early humans learned to use stone blades and fire to soften tough foods. Paradoxically, herbivorous primates have much larger canines than we do because their large teeth are used to signal aggression, a function that became unnecessary when we invented weapons.

children. Some scientists think it was the evolution of this system of "grandmothering" that extended our life expectancy and slowed the entire human life course as a side effect, delaying puberty well into the teens.

COMPLEX SOCIETY

Monkeys and apes have complex social lives and live in troops (tight-knit social groups). Many features common to monkey and ape societies are also true of human society. For instance, to get on in life, other primates have to climb a social ladder, exploiting a network of allegiances to get to the top. High-status animals attract the most mates and get first choice of the most important resources, such as food. Hierarchies and status are clearly also important in human society, and they provide us with similar benefits.

Understanding the complex relationships within a social group requires a certain type of intelligence, known to psychologists as Machiavellian intelligence. Primates are thought to use Machiavellian intelligence to keep track of their friends and enemies in the constantly changing social environment. The cleverest primates, the apes, even seem able to trick and deceive each other. To climb the social ladder, primates spend much of their time grooming each other's fur, an activity that seems to give them great pleasure. Humans are unusual among primates in that we do very little grooming, but then we have almost no hair to groom. According to one theory, conversation may serve the same purpose as grooming.

The need for Machiavellian intelligence may be one of the factors that led to what is arguably our most distinctive and important feature: a large brain.

BIG BRAINS

The organ that underlies most of our special abilities is the brain. Exactly what is so special about the human brain is largely a mystery. In terms of hardware (the cells and tissues of which it is made), it is much the same as any mammal's brain. In terms of size, though, it is unusually large.

Humans do not have the largest brains in the animal world; those of elephants and whales are much bigger. However, if the scaling effect of body size is taken into consideration, humans have the largest brains of any mammal. Our brains are not only huge, they are also expensive to run in that they consume a disproportionate amount of our energy: an adult's brain accounts for 2 per cent of body weight yet uses 20 per cent of calorie intake; the brain of a newborn baby uses 60 per cent of its energy intake.

The part of the brain that is thought to be responsible for our higher mental faculties is the cerebral cortex – the "grey matter" in the outer part of the cerebrum (the main area of the brain). In both humans and apes, the cerebral cortex is thrown into a series of deep folds that give the brain a very large surface area. However, humans have an especially large area of cortex: about four times as much as chimps. Neuroscientists know that damage to specific areas of this cortex can impair our higher mental functions, including language, decision making, memory, and emotional control – but as yet we have very little idea of how a healthy cortex manages to conjure up any of our unusual mental abilities.

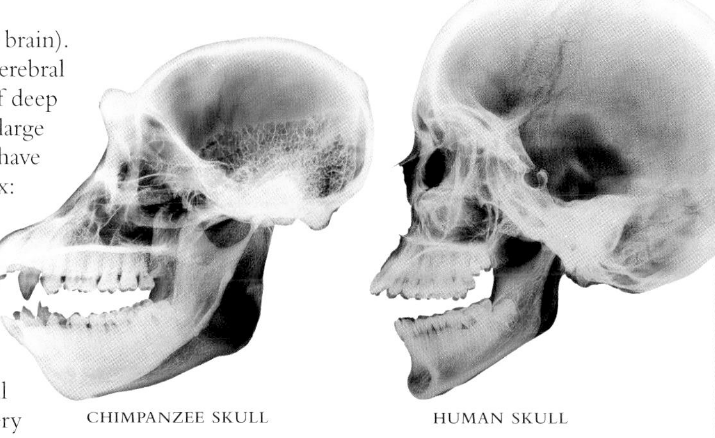

SKULLS AND BRAINS
As humans evolved, brains got bigger and jaws got smaller. Our jaws and teeth are smaller than those of chimpanzees because we use tools rather than teeth to prepare food. Our brains are three times larger, making our skull appear grotesquely swollen in comparison.

Scientists are not sure why our large brains evolved. The benefits seem obvious now, given the wealth of things we invent to make our lives easy. However, evolution cannot plan ahead, and our ancestors' brains increased in size hundreds of thousands of years ago. Since big brains consume much energy, there must have been an advantage to having them when we were living as foragers on the African savanna, but what was it?

One idea is that the evolution of the human brain was driven by the development of language. Language seems to have changed the structure of our brain: the speech areas make it asymmetrical because they are only found on one side, which has led to the human tendency to be right- or left-handed. The brains of other animals tend to be symmetrical, and it is likely that our early ancestors' brains were too. However, most experts think language actually appeared later than big brains – perhaps as recently as 50,000 years ago.

A more radical idea is that large brains are merely a side effect of the evolutionary process that increased our lifespan. Since brain tissue cannot regenerate, we need large brains just to provide the spare capacity that a longer life demands. This theory fails to account for the brain's voracious demand for energy, however. Our brains consume lots of energy because they are active; unused areas of "spare capacity" would be likely to lie dormant.

Another theory is that large brains evolved to make us sexually attractive. Charles Darwin was the first person to propose this idea. He suspected that sexual selection may have driven the evolution of intelligence and large brains in men in the same way it drove the evolution of gigantic tails in peacocks. One problem with this idea is that it accounts only for the intelligence of males, but men and women actually have the same average intelligence levels.

HIGH-DENSITY SOCIETY
Complex society may have been a driving force behind human intelligence. Today we live in cities of many millions, such as Shanghai in China.

It takes some ingenuity to visualize the shape of a blade in a lump of stone, so perhaps our ancestors' brains expanded as they honed their toolmaking skills. Surprisingly, evidence suggests this is not the case: prehistoric humans seem to have spent millions of years using the same old stone handaxes without innovation (although they may perhaps have been inventing ingenious tools from more perishable materials).

Then again, large brains may be related to our carnivorous diet. Animals that hunt for a living have large brains because they need to outwit their prey. Yet, although we eat more meat than other primates, other predators manage with much smaller brains than us, so hunting cannot be the only reason for human brain size.

An idea that has gained ground over recent years is that our large brains evolved for social reasons. Primates need cunning to understand the complexities of their social lives. Indeed, scientists have discovered a strong correlation between brain size in primates and the size of their social group. Our social groups are among the largest and most complex of any primate.

Perhaps the most convincing explanation for our intelligence, however, is that we evolved to fill what psychologists call a "cognitive niche". All species fill a niche that defines their way of life in an ecosystem. Our foraging ancestors were pioneers in using thought and knowledge to attain goals in the face of obstacles. Intelligence allowed them to think up countless new ways of finding food – by catching prey using traps, digging up plant roots with sticks, cracking open bones with rocks, and so on. As psychologist Steven Pinker has pointed out, "life for our ancestors was a camping trip that never ends, without the space blankets, Swiss army knives, and freeze-dried pasta".

TOOL USERS AND MAKERS

Many animals use tools (among them dolphins, sea otters, and vultures), and there are a few that make them (chimps and orangutans), but humans are unsurpassed in mastery of tools. Opposable (flexible) thumbs and superb manual dexterity all contribute to making us expert toolmakers. Toolmaking also demands a certain amount of brain power, especially insight. With insight and imagination, we are able to visualize designs in our head and solve problems before we set to work with our hands.

Our technical intelligence has also given us fire for cooking, buildings, transport, spacecraft, medicines, and many other wonders of the modern world. However, there is a dark side to our mastery of tools. The first stone tools made by our ancestors were stone blades, and they may have been used for violence against people as much as for hunting. Violence is one of the hallmarks of *Homo sapiens*; with the possible exception of chimpanzees, we are the only species known to engage in warfare.

THEORY OF MIND

Most animals see a stranger when they look in a mirror, but humans are among the few exceptions that recognize themselves. This rare ability hints at something special about our species: we each have a personal identity; a concept of the self. With this comes the ability to reflect on our own thoughts and feelings, and to visualize ourselves in imaginary situations. In turn, we can imagine what is going on in other people's minds by mentally putting ourselves in their shoes. Psychologists often refer to this phenomenon as a "theory of mind".

A theory of mind is very useful for a species such as ours, which lives in complex, dynamic social groups. Without it, we would not be able to deceive one another effectively, work out the devious motivations of our enemies, or predict how our friends and enemies might behave.

Chimpanzees may also be able to recognize themselves in mirrors, but does this mean they too have a theory of mind and the same kind of self-awareness as we do? Tantalizing anecdotes suggest they sometimes deceive each other, but as yet there is no unequivocal evidence for a theory of mind in any species besides our own.

Issue

Weapons of war

Humans are the only species that regularly goes to war, commits genocide, and has the capacity to self-destruct. War has probably always been an important driving force in the invention of new technology: some of the first human tools were weapons, and we probably used them against each other as much as for hunting. Our passion for war was also the driving force behind some of our greatest scientific achievements, from splitting the atom to landing men on the moon.

A TALKING APE

Language is arguably the greatest of human inventions. Other animals can communicate, but their efforts and abilities pale in comparison to our own. Vervet monkeys and meerkats have a vocabulary of alarm calls to warn each other of approaching predators, and even bees use a type of sign language to tell each other the way to nectar or pollen. However, other animal languages seem to lack one vital ingredient: grammar – a set of rules that enables words to be formed and strung together in a variety of different combinations, each with a distinct and unambiguous meaning.

The speed at which human language is both generated and interpreted is no less impressive than its infinite complexity. During normal speech, the human voicebox and mouth can generate up to around 25 separate speech sounds (roughly equivalent to letters of the alphabet) per second. Listeners can absorb this flood of data and effortlessly decode its meaning in an instant.

Language depends on the human ability to think symbolically. Words are essentially symbols for objects or concepts that need not be present or even physically real, for example people, food, emotions, and spirits. Symbolic thought also underlies another unique human creation: art. For tens of thousands of years, people have drawn paintings on rock walls, carved precious stones into ornaments, and adorned their bodies with tattoos and coloured dyes.

HUMAN CULTURE

Thanks largely to language, humans possess culture – the complex patterns of behaviour and knowledge that vary among societies and are handed down through the generations. Culture can adapt and learn; in doing so, it helped our species discover new ways of living as we spread around the globe. Through culture, society itself becomes a body of knowledge, obviating the need for each generation to relearn survival skills from scratch. It is through language that knowledge can be quickly shared and survival skills passed on.

Common to all the world's human cultures is the ability to form contracts and make deals. We remember social commitments and feel a sense of moral outrage and a desire for vengeance when betrayed. These are uniquely human traits; other animals have no concept of commitment or moral obligation, and as such, their ability to cooperate is limited. Humans build complex societies that are based on a division of labour. Social rules allow us to specialize in separate careers, yet live together and support each other by exchanging the fruits of our labour.

Key to the human ability to form contracts is our ability to remember the past and plan for the future. Mentally, we can travel forwards or back in time. Although other animals may have sharper memories than ourselves (squirrels and jays can remember the hundreds of sites in which they have buried food), it seems that humans are unique in that their memories are so clearly etched in time.

EARLY CAVE PAINTING

THE CREATION OF ART

Art and symbolic thinking may have existed for as long as our species. Even the oldest cave paintings show a sense of perspective comparable with that of the greatest modern works of art.

LEONARDO DA VINCI'S *MONA LISA*

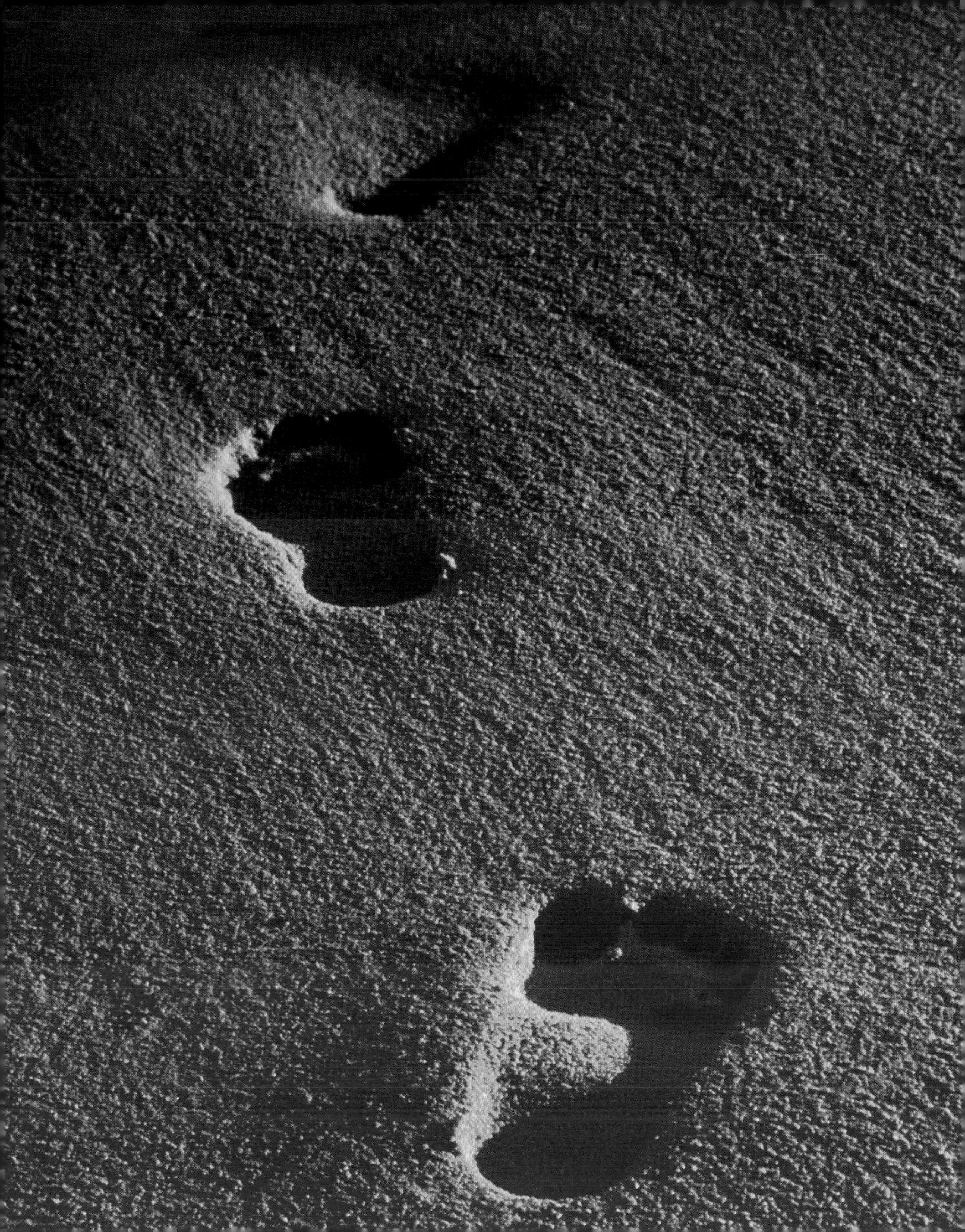

ORIGINS

ORIGINS

We must look to the past to see how far the human body, mind, and culture have evolved. The story winds through 6 million years, most of which are shrouded in mystery. Historians look back a few thousand years, to the dawn of writing and recorded history. Archaeologists peer a little farther back, digging through the litter of ancient civilizations and disinterring the dead. Beyond that, we have just faint glimpses of our past: the occasional bone or tooth, a scattering of stone tools, and a few clues from our DNA.

Some 6 million years ago, Africa's forests were home to an ape that probably looked quite similar to a chimpanzee. Like a chimp, it was an agile climber and spent a lot of time in the trees, but could perhaps take a few steps on two feet. It may also have mastered a few tools. By approximately 5 million years ago, the species had split in two. One group stayed in the tropical African forests, giving rise to the chimp and its close cousin the bonobo. The other adapted to life on land, gradually starting to walk upright and spreading to the savanna.

PLANET HOMINID

This ground-dweller was the first member of a branch of the ape family that has been traditionally known as hominids, a group that ultimately came to include ourselves. However, classifications are being amended to refer to all great apes as "hominids". As a result, our ancestors from after the time of the split with the chimp line are now frequently termed "hominins". Here, the term "hominids" will continue to be used when referring specifically to our walking ancestors, and not to the other great apes.

Palaeontologists have now laid to rest the theory that big brains were responsible for driving us to walk upright in order to leave our hands free for making tools. In fact, our ancestors were walking on two feet by 4 or 5 million years ago, when their brains were no bigger than those of chimps. Even after their brains expanded, our ancestors continued to spend millennia hammering out the same crude tools with little sign of intelligence. The full impact of the human brain did not make itself apparent until around 40,000 years ago. By that time, though, a full 99 per cent of our history was behind us.

SOLE SURVIVOR

Many years ago, scientists saw each species as an ancestor of our own. It seemed logical to organize them in a linear sequence, each one slightly more "human" than the last. With the discovery of more fossils, a very different picture has emerged. Human evolution looks like a bush made up of a maze of dead ends, and working out how all the species relate to each other is very difficult.

One of the most sobering discoveries is that our story is one of failure and extinction. All of our ancestors' contemporaries perished, leaving *Homo sapiens* as the sole survivor. Evidence suggests that our species very nearly met the same fate when population crashes significantly reduced its numbers. It may therefore be a matter of luck that we are here at all. Fossils reveal that our "cousins" survived

Issue

Where do we come from?

Some cultures believe in a single deity that created the world suddenly: many Christians believe God created Adam and Eve just 6,000 years ago. Other societies think the world developed organically, like a fetus in the womb, or from "parents" symbolized by the sky and earth. Aboriginal Australians believe the world was created by mythological ancestors during the "dreamtime". In Judaeo-Christian mythology, special powers made us superior to plants and other animals; in some cultures, though, these entities embody the spirits of our ancestors.

ABORIGINAL DREAMTIME FIGURES

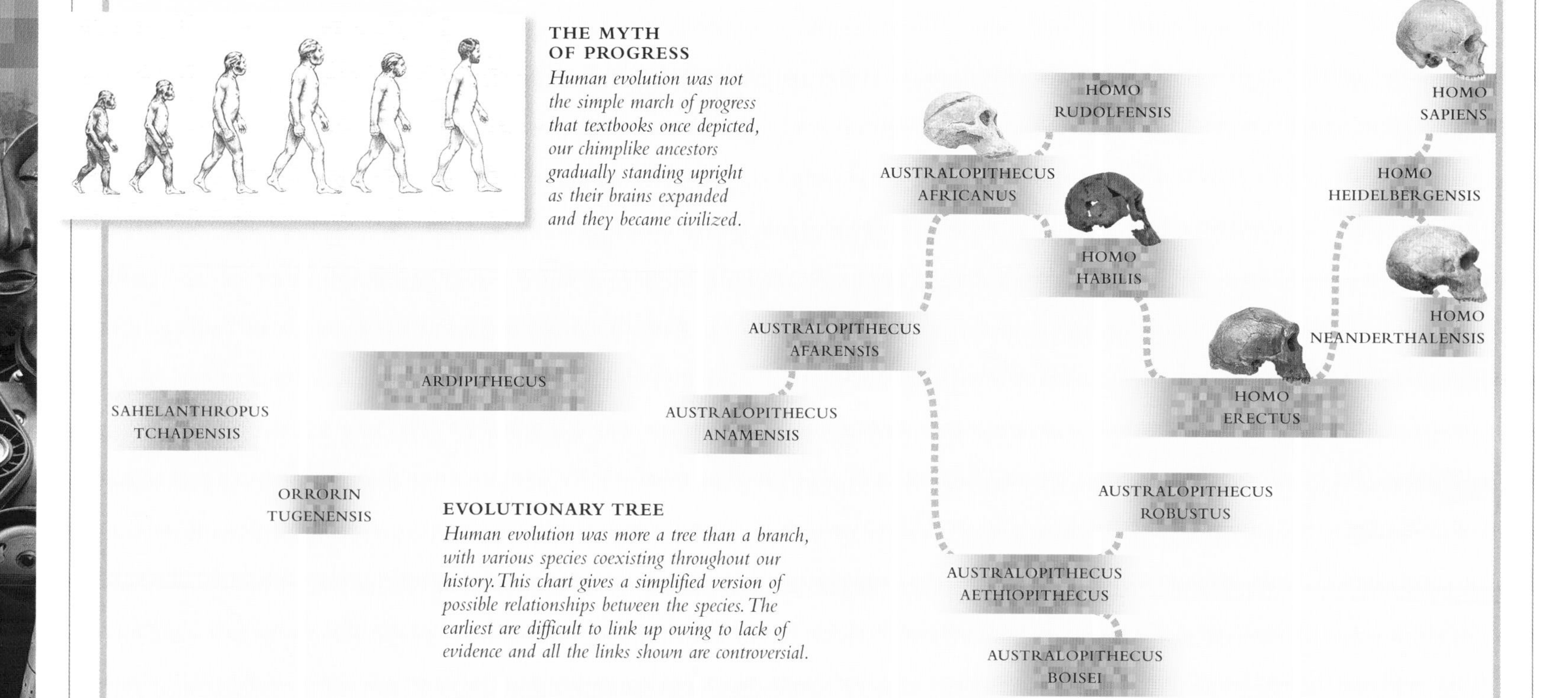

THE MYTH OF PROGRESS
Human evolution was not the simple march of progress that textbooks once depicted, our chimplike ancestors gradually standing upright as their brains expanded and they became civilized.

EVOLUTIONARY TREE
Human evolution was more a tree than a branch, with various species coexisting throughout our history. This chart gives a simplified version of possible relationships between the species. The earliest are difficult to link up owing to lack of evidence and all the links shown are controversial.

until as recently as 30,000 years ago, and that small pockets may in fact have clung on even longer. Almost every culture seems to have stories about mythical apemen, from the Sasquatch in North America to the Yeti in the Himalayas, the Alux in Central America, and the Orang Pendek in Sumatra. Perhaps these myths originated as Chinese whispers from a distant past when our hominid cousins still lived.

SO WHY BIG BRAINS?

A traditional explanation for why we evolved such large brains focuses on technology: perhaps they evolved to help us devise complex weapons. With these, we unlocked a valuable new source of food – meat – which provided the protein and calories to make our brains even bigger. Yet perhaps stone tools dominate our thinking simply because so many examples have been found.

The probable benefits of big brains were complex language and social behaviour. Recently, the role of social interaction in evolution has been given more emphasis. In human society, success depends on the ability to exploit a web of social relationships. This skill may have been just as important as the ability to sharpen a rock or throw a spear.

LEAPING FORWARD

Human history is all about quantum leaps. For thousands, even millions, of years, nothing much happens; then some discovery or twist of fate propels humanity forward.

One such leap showed itself around 40,000 years ago, when culture, art, and sophisticated tools made an appearance. In Europe, this cultural revolution is associated with the Cro-Magnons, whose paintings adorn the caves of southern France and northern Spain. Some experts claim the trigger was the origin of language and consciousness, but this remains an area of debate. The hunter-gatherer lifestyle remained our species' only way of life until around 10,000 years ago, when another leap occurred.

CIVILIZATION DAWNS

As the last ice age drew to a close, people all over the world learned to domesticate and rear animals, and to cultivate plants. The invention of agriculture set in motion events that were to transform human society. Instead of roaming, people settled down. A good food supply led to a growth in populations and the emergence of towns. City life led to a greater division of labour and increased innovation.

RECORDED HISTORY
Written language, such as this Sabaean script that dates from around 500BC, sprang from the emergence of civilization.

However, evidence suggests the first farmers were less healthy than their nomadic ancestors. They lived in crowded, insanitary conditions and their diet was dominated by just a few staple grains. Social divisions also appeared: a small number of landowners lived in luxury, while most people toiled in the fields.

THE AGE OF REASON

Writing is the sacred text of our human history. Written language was initially used only to draw up transactions and contracts. Later, however, it began to be used to record sacred knowledge and mythology, and by the time of ancient Greece, it was being used to store and communicate knowledge for its own sake.

This pure love of knowledge has continued to feature in intellectual movements, such as the Renaissance. This began in Italy in the 14th century and continued in Europe until around the middle of the 17th century. It was a time when action and reason started to challenge religious and contemplative life. The Renaissance was an era of exploration and discovery in every sense. It pushed back the boundaries of art, geography, music, science, and thought. Writing became a tool for recording and sharing new knowledge, be this scientific, artistic, or political.

REASON BRINGS REVOLUTION
Increased interest in philosophy and politics not only led to cultural advances but gave the populace the tools with which to challenge its masters, as occurred in the French Revolution.

DIVISION AND UNITY

For most of history, groups of *Homo sapiens* did not share knowledge; it was passed from one generation to the next or discovered anew. When groups met it was often to fight – something humans seem predisposed to do. As technology develops, this seemingly innate urge could actually prove to be a threat to humanity. The more we learn about our world, the more we seem to be capable of destroying it. Today we live in a global village of 6.4 billion people and rising; all our knowledge is pooled; many of us speak the same language; and every major town is connected to the internet. The cultures that once divided us may be breaking down and merging. If the past is anything to go by, society will continue to change at breakneck speed. Only time will tell where this will take us.

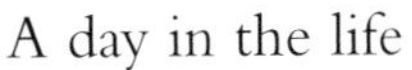

Fact

A day in the life

Imagine human history as a 24-hour clock, starting with the split from the other apes. We lived as hunter-gatherers until very nearly midnight. Suddenly, at 23:58, agriculture and towns appeared. We invented the wheel and built the pyramids at just over a minute to 12; the first Olympics took place at 40 seconds to; man walked on the moon at 3 seconds to; and the internet came about at a tenth of a second to. All of our greatest achievements have only just happened and the pace is accelerating.

Fact

History repeating itself

The cradle of human evolution was the savanna of tropical Africa, where our ancestors survived for millions of years by gathering food and hunting. The legacy of this period is with us today. Despite thousands of years spent living all over the world, we remain tropical animals, dependent on warm clothes and heating in cold areas. Our food also reflects grassland origins: we choose grass seeds, in the form of wheat and rice, as our staple. And just as our ancestors survived best in small bands, we still sort ourselves into small communities, even in the most densely populated of cities.

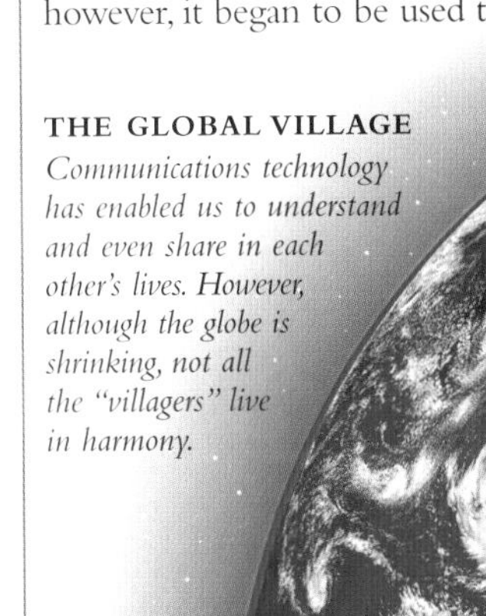

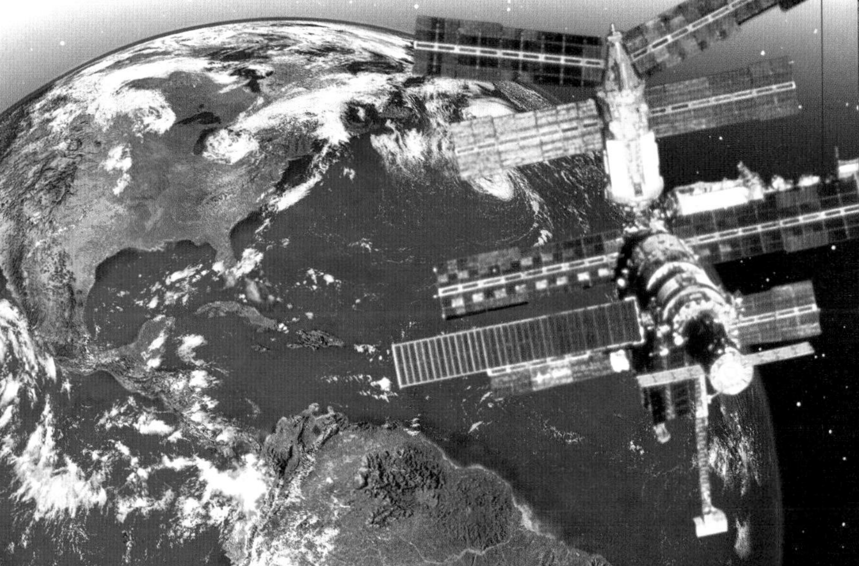

THE GLOBAL VILLAGE
Communications technology has enabled us to understand and even share in each other's lives. However, although the globe is shrinking, not all the "villagers" live in harmony.

FIRST STEPS

Around 6 million years ago, a momentous event occurred in the forests of Africa. An ancient species of ape left the safety of the trees and began to live on the ground. Exactly how, when, and why this species made the transition remain shrouded in mystery. Perhaps droughts dried out the rainforest, turning it into savanna. Or perhaps the land flooded, forming food-rich lagoons that tempted animals out of the trees. We may never know. One thing, however, we do know for sure: one pioneering ape adapted well to life on the ground. Instead of walking on its knuckles, it took to balancing precariously on its hind limbs, with its forelimbs dangling at its sides. It was the first step in an extraordinary process that was to transform apes into human beings.

AN UNUSUAL SOLUTION
Most big apes cope with life on the ground by knuckle-walking. Our ancestors were different: they stood up and walked.

A FAMILY IN DECLINE

Apart from *Homo sapiens*, the apes are a family in decline. Apes probably reached their peak around 20 million years ago, when at least 60 species inhabited the tropical forests of Africa and southern Asia. Around 10 million years ago, the apes started to disappear, and monkeys began taking their place. For millions of years, the planet has been getting cooler and drier, with the vast belt of rainforest once found in the tropics turning into sparser woodland or open grassland. The apes were fruit specialists, dependent on rainforest for their main source of food. So, as the forests dwindled, the apes disappeared. Monkeys learned to exploit the nuts and seeds of their new home. They had a secret weapon: the ability to digest unripe fruit, allowing them to steal the apes' food before it was edible. Around 5 million years ago, the Earth went through an especially severe bout of cooling and drying. In East Africa, the effect was heightened by slowly drifting continental plates generating immense pressure deep underground. The African continent tore open, producing a gash of monumental proportions: the Great Rift Valley. Into this landscape walked a new species with a secret weapon of its own: the ability to walk on two feet.

THE GREAT RIFT VALLEY
About 5 million years ago, the East African landscape began to pull apart, replacing dense jungle with the mosaic of grassland, desert, woodland, and rivers that exists today.

Issue

Why did we walk?

Perhaps standing enabled our ancestors to see over tall savanna grass to spot predators and prey, or perhaps they needed to be able to carry things. One idea is that an upright stance reduced exposure to the sun, allowing them to stay active during the midday heat. A controversial theory is that we became two-legged during an aquatic phase in our evolution. Animals that could wade in shallow lagoons would have found a rich supply of food.

ON TWO FEET

Whatever tipped the balance towards a bipedal stance, becoming upright had its benefits. Walking on two legs uses less energy than knuckle-walking; more importantly, it leaves the hands free for many other purposes, including making tools, carrying things, gathering food, and throwing weapons. *Ardipithecus* is the oldest-known certain hominid (upright ape). Two distinct species have been found: *Ardipithecus kadabba* existed around 5.8 million years ago and *Ardipithecus ramidus* around 4.4 million years ago. Based on the few fossils found, *Ardipithecus* had a mixture of hominid and ape characteristics. It had the short canine teeth typical of hominids, but its molars were apelike in shape and its incisors large – more like a chimp than a human. The skull bones found suggest the skull rested on top of the spine, making *Ardipithecus* bipedal, but this has not been confirmed. Its arms appear large and powerful, so it was probably still a good climber. Maybe it spent its days on the ground but returned to the treetops to sleep, like chimps and gorillas do. Evidence suggests that *Ardipithecus* inhabited a woodland environment, not the grassy savanna previously believed to be home to the earliest hominids.

EARLY HOMINID

ANCIENT APE

BIPEDAL STANCE
On leaving the trees, our ancestors had to adapt to life on the ground. They gradually became proficient at walking on two legs.

7 million years ago
Sahelanthropus (see The Chad skull, opposite page) in existence

5.8 million years ago
Earliest evidence of *Ardipithecus kadabba*, the first known hominid

5–6 million years ago
Human branch had separated from chimp branch of ape family tree according to some genetic estimates

7,000,000 | 6,000,000 | 5,000,000

6 million years ago
Leg bones dated to this time claimed to be the earliest evidence of upright walking on two legs

5 million years ago
The Earth begins to go through severe bouts of cooling and drying, with rifting in northeast Africa

EARLY HUMANS

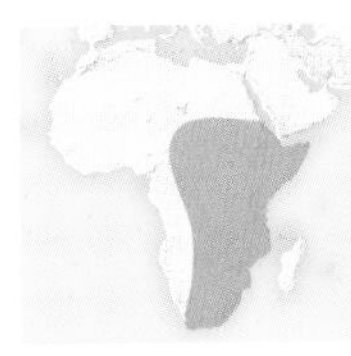

WHERE THEY LIVED
Australopithecus is known to have inhabited much of eastern, southern, and central Africa.

Australopithecus was a common and successful type of early hominid judging by the many fossils found. Several species spanned 3 million years, and these species divided into two main types: gracile and robust. Gracile australopithecines are likely to have been our ancestors and died out around 2.5 million years ago; robust australopithecines were an evolutionary dead end but survived later than the graciles. Both types of australopithecine would have looked a bit like upright chimps: they were only about 1–1.5m (3–5ft) tall, lacked the athletic build of later hominids, and had flat nasal openings. The shape of fossilized spinal, pelvic, and leg bones shows they could walk upright, as do the tracks of footprints found in a stretch of 3.65-million-year-old rock in Laetoli, Tanzania (*see below*). However, their short legs and long arms suggest they may only have been occasional walkers who spent much of their time sitting or climbing. Gracile species had smaller back teeth, and therefore probably subsisted on a varied diet of fruit, insects, seeds, roots, and maybe meat. The bigger jaws and back teeth of robust australopithecines point to a diet of coarse grassland vegetation. There is no real evidence that australopithecines could make stone tools. They perhaps made simple tools out of twigs, but may have lacked the intelligence to create anything more sophisticated.

AUSTRALOPITHECINE SKULL
The skull of Australopithecus *was shaped quite differently from ours, and its brain was only about the size of a chimp's. Shown here is the skull of* Australopithecus africanus, *a gracile australopithecine.*

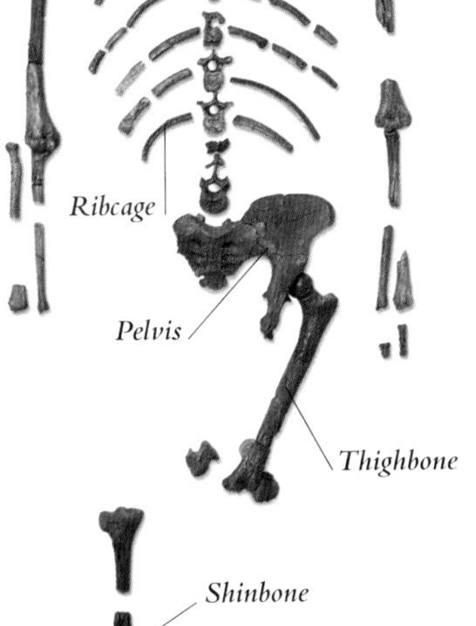

FAMOUS SKELETON
Almost half of this female Australopithecus afarensis *(gracile) skeleton survived its 3.2 million years of burial. "Lucy" was adult, but she stood only 1.1m (3ft) tall.*

EARLY FOOTPRINTS
These tracks, found in fossilized volcanic ash in Tanzania, may have been made by a family of three.

A SUCCESSFUL HOMINID
Judging from the many fossils found, Australopithecus *was a common hominid, its several species spanning around 3 million years.*

Issue

The Chad skull

From the front, this skull found in Chad in 2001 (named *Sahelanthropus*) is said to look more advanced than *Ardipithecus* or early *Australopithecus*. Yet it is 6–7 million years old, predating all known hominids and the hominid–chimp split. If it is a hominid, *Ardipithecus* and *Australopithecus* may be evolutionary dead ends; the split would also have to be pushed back or chimps redefined as hominids. We still do not know where *Sahelanthropus* sits on the ape evolutionary tree.

4.4 million years ago
Ardipithecus ramidus believed to be in existence

4.2 million years ago
Limited remains of bipedal hominid (*Australopithecus anamensis*) dated to this time found by Lake Turkana, Kenya

4,000,000

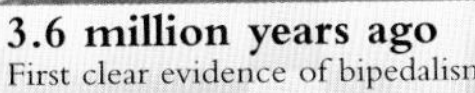

3.6 million years ago
First clear evidence of bipedalism: *Australopithecus afarensis* footprints found at Laetoli in Tanzania

3.2 million years ago
"Lucy" (*Australopithecus afarensis*) in existence; skeleton found at Hadar in Ethiopia

3 million years ago
Australopithecus africanus, notable for the powerful build of its upper body, in existence

3,000,000

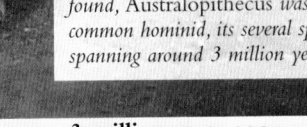

2.6 million years ago
Australopithecus aethiopicus in existence; the earliest finds of stone tools also date to this time

TOOLMAKERS

The arrival of *Homo*, the human genus (group of species), between 2 and 2.5 million years ago marks a major turning point in the story of our evolution. Apelike hominids such as *Australopithecus* had flourished, and continued to flourish, for millions of years with little change. With *Homo* there was a sudden leap in brain size, a dramatic change in anatomy, and the beginnings of stone-age technology. Perhaps the emergence of *Homo* also marked the first glimmerings of language, culture, and a social structure based on monogamous families. The origins of *Homo* are unclear. It may have evolved from one of the gracile australopithecines in Africa, but exactly where and when remain shrouded in mystery.

THE HANDYMAN

Homo habilis ("handyman") marks a turning point in human evolution: it could make stone tools. Humans have a lack of natural weapons such as sharp claws or teeth, but stone tools allowed our ancestors to become more carnivorous. The tools of *Homo habilis* were simple and included small, sharp flakes of rock, probably used as blades or scrapers for cutting hides and butchering carcasses. Some experts think *Homo habilis* scavenged the leftovers of other predators rather than hunting live prey. Many of Africa's herbivores would have been too fast and strong to take on, but a gang of rock-throwing hominids could have driven cheetahs from their meals. The brain of *Homo habilis* probably consumed a disproportionate amount of energy. Stone tools unlocked the rich source of calories in meat necessary to power this hungry organ and sustain its expansion over the next 2 million years. A more versatile diet may also have helped to liberate our ancestors from their habitat.

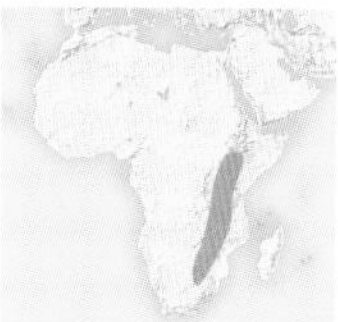

WHERE AND WHEN?
Homo habilis is thought to have lived in East Africa between 2.3 and 1.6 million years ago.

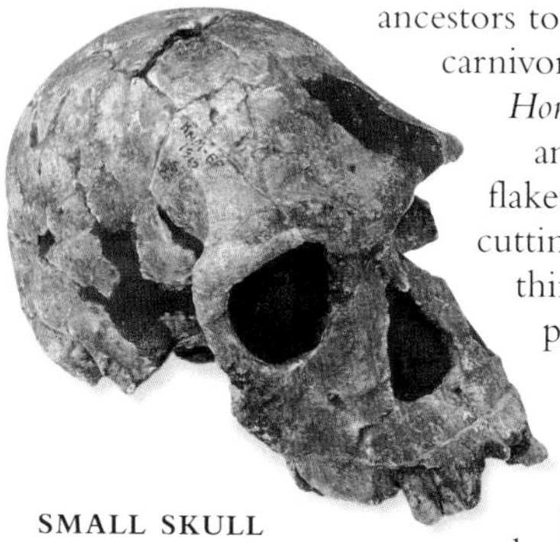

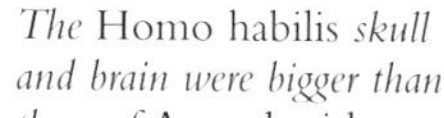

SMALL SKULL
The Homo habilis skull and brain were bigger than those of Australopithecus, but its brain was only half as big as a modern human's.

FIRST TOOLS
Sharp flakes of rock were used as blades or scrapers. They were struck from cobbles that may also have been used for cracking nuts or releasing marrow from bones.

HUMAN GENUS
Homo habilis is one of the earliest species assigned to the human genus, and one of the first species known to have made stone tools.

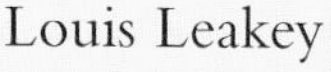

Profile

Louis Leakey

Born in Kenya, Louis Leakey (1903–1972) always believed Africa was the cradle of human evolution, although many disagreed. He began fossil-hunting in East Africa's Great Rift Valley, and in 1959, he and his wife Mary discovered the oldest-known stone tools. Leakey's conviction that humans originated in East Africa led to a string of major fossil finds, including those of ancient apes. Leakey's son Jonathan found *Homo habilis*, his son Richard the Turkana boy (*Homo erectus*), and his wife the Laetoli footprints (*see* p25).

2.3 million years ago
Homo habilis appears

2,500,000

2 million years ago
Australopithecus robustus comes into existence

1.8 million years ago
Homo erectus, the first hominid to leave Africa, first emerges

2,000,000

1.7 million years ago
Earliest evidence of hominids in Asia

1.6 million years ago
Homo habilis dies out

THE FIRST TO LEAVE AFRICA

Homo erectus ("upright man") first appeared in East Africa around 1.8 million years ago. *Homo erectus* was the first hominid to leave Africa and seems to have spread across the Old World with astonishing speed: by 1.7–1.8 million years ago it had reached Georgia, and by 1.6 million years ago it was in Java. Much of what we know about *Homo erectus* comes from a single individual: the Turkana boy. His 1.5-million-year-old, almost-complete skeleton was found in 1984 near Lake Turkana in Kenya. The Turkana boy was aged 8–11 when he died face down in a marsh that buried him before scavengers could destroy the carcass. He looked radically different from *Australopithecus*. Despite his age, he stood 1.6m (5ft 3in) tall; an adult may have been well over 1.8m (6ft). His long legs and narrow pelvis show he was as upright and athletic as us, but his sturdier bones suggest a more muscular build. His face projected less than that of *Australopithecus*, and he had a nose instead of flat nasal openings. Yet a prominent bony ridge jutted out above the eyes and would have given him a glowering expression. His brain was still far smaller than the modern average, but other physical measurements suggest the late puberty, or long childhood, characteristic of modern humans had started to evolve. Our long childhood is linked to the fact that we are born earlier in our development than other animals. Human infants are thus helpless and depend entirely on their parents, who tend to form long-term couples in which to look after them. The apparent late puberty of *Homo erectus* suggests this social feature had already begun to emerge.

UPRIGHT MAN
Homo erectus *was tall with long limbs, helping the body to shed heat. It is likely that he had bare skin to aid sweating, and he would have been black for sun protection.*

ON THE MOVE
Homo erectus *spread out of Africa around the globe, possibly surviving until as late as 100,000 years ago in China and Java.*

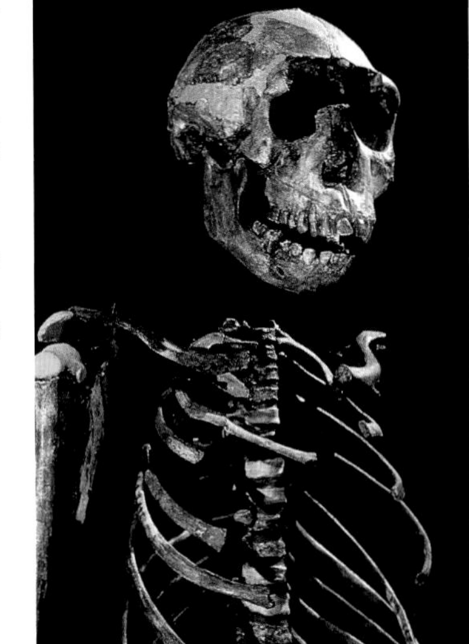

TURKANA BOY
His barrel-shaped ribcage indicates a smaller abdomen and shorter intestine than Australopithecus, *a sign he ate more meat.*

INTELLECT AND INNOVATION

The still-small brain of *Homo erectus* may reflect a lack of intellect or simply a short lifespan. The shape of its skull base suggests it had a lower larynx than an ape, so could talk. But the Turkana boy's spinal cord was much narrower in the chest than ours and his chest muscles were unlikely to have had the nerve connections needed for true language. The hand axes made by *Homo erectus* were multifunctional tools probably used to skin and butcher animals. They were made with skill, yet hardly changed in over a million years, which suggests a mind very different from our own. It is as if this species was driven to make the same tools again and again. In 1997, 840,000-year-old stone tools, possibly those of *Homo erectus*, were found on an Indonesian island that may never have been joined to the mainland. Some therefore believe *Homo erectus* could build rafts from wood or bamboo and sail across the sea. Palaeontologists have also found evidence of a new type of sociality: one individual survived for months after being crippled, so someone may have been caring for her. A darker picture of early humans comes from Europe 800,000 years ago. *Erectus*-like creatures left behind hominid bones bearing scratch marks from being defleshed using tools.

EVER-INCREASING SIZE
The brain of Homo erectus *was somewhat bigger than that of* Homo habilis; *a more complex social life could be the explanation.*

MORE ADVANCED TOOLS
More sophisticated than the cobbles and flakes of Homo habilis, *the hand axe of* Homo erectus *was worked into a teardrop-shaped cutting implement by chipping it on both faces.*

SIGNS OF CANNIBALISM?
Hominid bones found at Burgos in northern Spain bear scratch marks made when the flesh was removed – possibly a sign of cannibalism or ritual defleshing of the dead.

Scratch marks

Issue

Discovery of fire

Hearths first appear in the fossil record 250,000 years ago, and ash deposits in China are seen by some as the remains of 400,000-year-old hearths. Yet *Homo erectus* may actually have used fire 1.5 million years ago. Cooking turned indigestible plant matter into energy-rich brain fuel, so could explain the species' small teeth and intestines. It may even have transformed society: food now had to be gathered, carried, and prepared, and a female would have been at risk from thieves. The need for male protection could be a possible origin of the male–female bond.

1.5 million years ago
"Turkana Boy" (*Homo erectus*) in existence; hand axe developed; earliest evidence of fire, southern Africa

800,000 years ago
Oldest evidence of human life in Europe, near Burgos in northern Spain

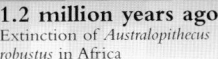

1,500,000

1,000,000

1.2 million years ago
Extinction of *Australopithecus robustus* in Africa

THE HUNTER-GATHERER

There is evidence of some European life before 600,000 years ago: fossils of *Homo antecessor*, a creature similar to *Homo erectus* that has been dated at 800,000 years old, have been found in northern Spain. However, we only have clear evidence that Europe was well populated from 600,000 years ago. Early Europeans were without doubt specialized hunters: both *Homo heidelbergensis* and *Homo neanderthalensis* are known to have crafted hunting tools, and some evidence suggests that they may even have hunted large animals at very close range.

A WAY OF LIFE
It is likely that both Homo heidelbergensis *and* Homo neanderthalensis *hunted deer and other large animals.*

Issue

The Pit of Bones

The 400,000-year-old bones of about 32 *Homo heidelbergensis* bodies were found in a deep pothole in northern Spain. The site is thought by some to be a burial chamber with spiritual meaning. Ritualistic burial implies the dawn of sophisticated consciousness, so this would give the site huge significance. Others are less convinced, suspecting the bones slid into the pit or were dragged there by animals.

MAN THE HUNTER

Homo heidelbergensis came into existence around 600,000 years ago. Although experts are unsure whether *Homo erectus* was more of a hunter or a scavenger, there is little doubt that *Homo heidelbergensis* was a skilled hunter. At Boxgrove, England, animal bones and stone tools have been found at what seems to be a prehistoric butchery site. Most of the animals were large game, such as deer and rhinos, and their bones bear scratch marks made by stone tools. These marks lie under the tooth marks of scavenging animals, showing that the hominids got to the prey first. More evidence of hunting comes from Shöningen, Germany, where 400,000-year-old hardwood spears have been found. The spear ends appear to have been split, possibly to hold stone spearheads.

INHABITED EUROPE
Homo heidelbergensis *lived in Europe, as well as in parts of Asia and Africa.*

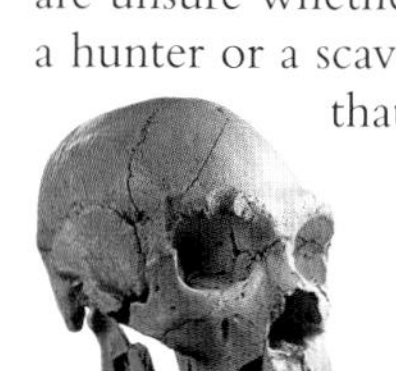

SKULL

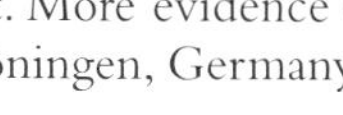

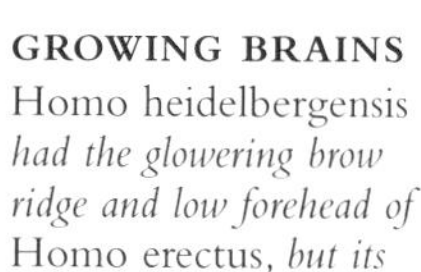

RECONSTRUCTION

GROWING BRAINS
Homo heidelbergensis *had the glowering brow ridge and low forehead of* Homo erectus*, but its brain was bigger.*

ICE-AGE SURVIVORS

Few hominids have caused as much controversy as *Homo neanderthalensis*. Wearing a hat and scarf, a Neanderthal might attract little attention in the street today. Yet with his or her face uncovered, it would be different. A bulbous nose made the centre of the face jut out and, instead of a forehead, there was a prominent brow ridge overshadowing deep-set eyes. The Neanderthal chin was receding, wide cheekbones flared back on each side of the face, and the front teeth were huge. People tend to think of Neanderthals as primitive cavemen, stooped and as hairy as apes. This distorted view of our close cousin is largely the result of the Old Man of Chapelle, a fossil found in 1908. Scientists emphasized his curved spine, stooped posture, apelike feet, and limited intellect. In the 1950s, however, the Old Man of Chapelle's curved spine was found to be the result of arthritis and his brain bigger than the modern average. Neanderthals were well adapted for surviving the cold Ice-Age winters: their short, stocky build helped to conserve heat.

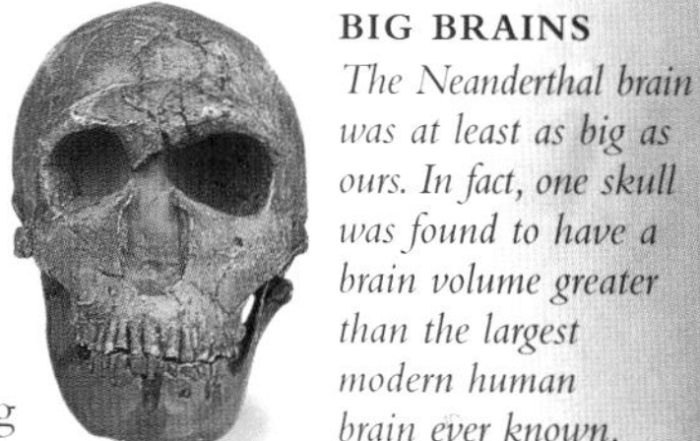

BIG BRAINS
The Neanderthal brain was at least as big as ours. In fact, one skull was found to have a brain volume greater than the largest modern human brain ever known.

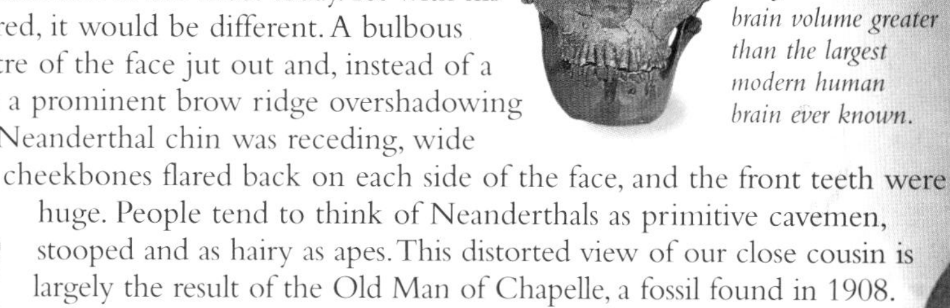

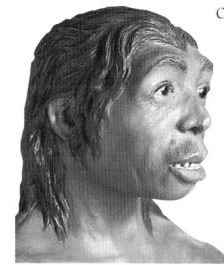

SKIN VARIATION
Most experts now believe that Neanderthal skin was hairless and varied from pale to dark.

COLD CLIMES
Neanderthals lived in Ice-Age Europe and western Asia between 250,000 and 30,000 years ago.

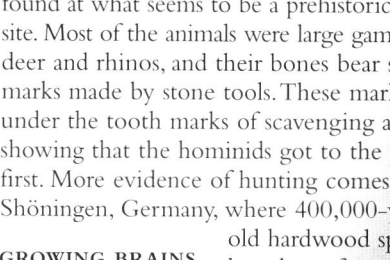

NEANDERTHAL MAN
Perceptions of our close cousin Homo neanderthalensis *alternate between hairy savage and sophisticated hunter-gatherer.*

600,000 years ago
Emergence of *Homo heidelbergensis* in Europe, northern Africa, and western Asia

400,000 years ago
Oldest preserved wooden spears found in Shöningen, Germany, and Clacton, UK

600,000 | 500,000 | 400,000

400,000 years ago
The Pit of Bones ("*La Sima de los Huesos*") in the Atapuerca region of northern Spain dated to this time

GRIP COMPARISON
Neanderthal hands (see the fingertip bone near right) may have given a less subtle grip than those of Homo sapiens *(fingertip bone, far right).*

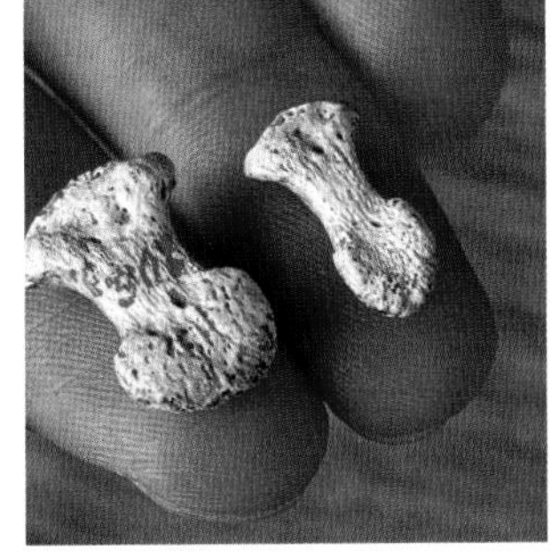

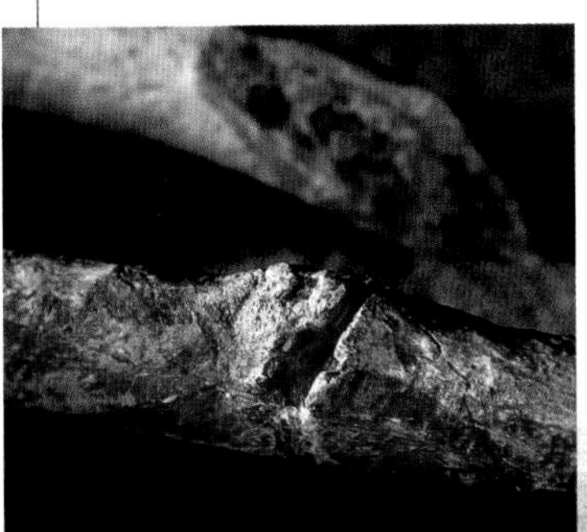

BROKEN BONES
Fractures and other traumatic injuries are almost universal among Neanderthal fossils, especially on upper body bones such as the ribs shown left.

A LIFE OF VIOLENCE

Neanderthals clearly took care of the elderly or infirm. At Shanidar Cave in Iraq, archaeologists found the skeleton of a 40-year-old man who had survived for several years with a catalogue of injuries and deformities. One side of his face was so badly crushed that he must have been partially blind, yet the wound had healed over. He had crippling arthritis in his right leg, healed fractures in a foot, and a withered arm. Any one of these afflictions could have killed him if he had lived alone, but with a social group to support him, he survived. The Shanidar skeleton's injuries bear witness to another feature of the Neanderthals – their violent lives. A likely explanation for the fractures commonplace among Neanderthal fossils is that they hunted big game at close range, using brute force and hand-held weapons to overpower animals as big as bison. Some experts think they may even have leaped on their prey. The price the Neanderthals paid for this physically demanding way of life was frequent injury and a life expectancy of only 30–40 years. Neanderthal tools were more advanced than those of earlier hominids and, judging from the pattern of wear on the Neanderthals' large front teeth, they used their mouths to hold meat while hacking at it with knives – much as Arctic people eat reindeer meat today. The Neanderthals also had fire: they built crude hearths and left layers of compressed ash on the floors of their caves. For all their skill with stone blades and fire, some experts think that Neanderthals were somewhat ham-fisted compared to modern humans, which may be why they rarely made wooden shafts for spears and knives.

BURIAL CEREMONY
Some Neanderthal graves contain animal bones, possibly left as part of burial rituals. This reconstruction shows a Neanderthal burial being performed.

MORE ADVANCED TOOLS
Neanderthals replaced bulky hand axes with smaller implements, such as flakes of rock retouched along the edges to make scrapers or saw-tooth cutters.

WHAT HAPPENED TO THE NEANDERTHALS?

The greatest controversy around the Neanderthals is why they disappeared. They flourished for more than 200,000 years, endured countless Ice-Age winters, and became skilled hunters of some of the biggest animals on the planet. Yet, in the end, something drove them to extinction; it was probably us. Modern humans reached Europe about 40,000 years ago – just 10,000 years before the last Neanderthals perished. It seems the two could not coexist, but exactly why remains a mystery. Perhaps our ancestors waged war on the Neanderthals or hunted them for food. Or maybe the Neanderthals succumbed to a disease brought by the invaders. In recent years, however, another theory has been gaining ground: that modern humans had some kind of cultural or intellectual advantage over the Neanderthals that gave them a competitive edge, especially when facing increasingly unstable climates. Our arrival in Europe is accompanied by the emergence of symbolic thought, art, sophisticated new tools made from bone and antler, and more effective clothing to keep out the cold. We brought with us a complex culture and formed networks of social contacts with neighbouring groups, fostering trade and the flow of information. All of this points to the final emergence of the modern human mind and, according to some experts, the flowering of language. The Neanderthals simply could not compete for Europe's diminishing resources in the unstable climates of the time.

Fact

Neanderthal DNA

In 1997, scientists sequenced a length of DNA from a Neanderthal bone. At the time, some experts believed that *Homo sapiens* had evolved from *Homo erectus* and *neanderthalensis*. In fact, Neanderthal DNA was found to be so different from ours that our most recent common ancestor could have lived no later than 500,000 years ago. The Neanderthals were not our ancestors, but a sister species we replaced as we spread out of Africa.

UNPICKING THE PAST
Many thousands of people around the world are involved in the search for evidence of where, when, and how the Neanderthals lived and what happened to them.

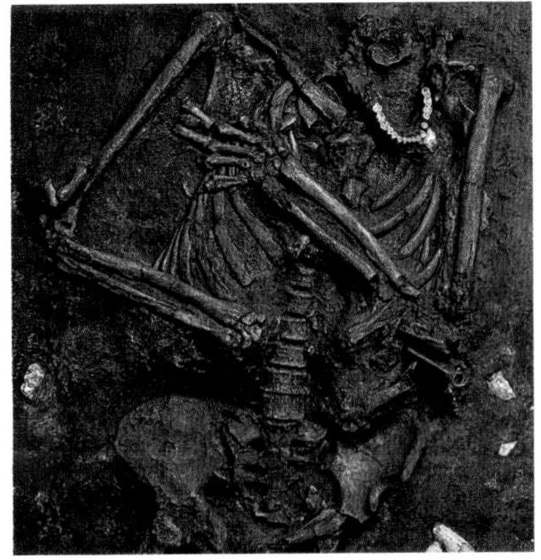

BURIED EVIDENCE
So much is known about the Neanderthals because they buried their dead, leaving many skeletons like this one for palaeontologists to unearth.

260,000 years ago
Possible earliest *Homo sapiens* in Africa

300,000

250,000 years ago
Homo neanderthalensis emerges and exists in ice-age Europe and western Asia for the next 220,000 years

160,000 years ago
Definite earliest *Homo sapiens* in Africa

200,000

MODERN HUMANS

Homo sapiens evolved 150,000 to 200,000 years ago in Africa. At first, our species led a similar life to other hominids, but modern behaviour began to evolve. From around 60,000 years ago, *Homo sapiens* swept out of Africa in waves, spreading across the planet and replacing the natives to become the only hominid on Earth. A cultural revolution occurred 40,000 years ago, with the appearance of Cro-Magnons (a type of *Homo sapiens*) in Europe. They were fully modern in their anatomy, society, and behaviour, and had a culture so sophisticated that some call this stage the "Great Leap Forward". In a few thousand years, there was more innovation than in the previous 6 million – a spectacular flowering of art, music, religion, mythology, trade, clothes, houses, and tools.

CRO-MAGNON SKELETON
This amazingly intact specimen shows how similar the anatomy of Cro-Magnons was to our own. They were also fully modern in their society and behaviour.

COMMON ANCESTORS

"NUCLEAR ADAM"
Studies of the Y chromosome (right, alongside an X) provide a detailed picture of our ancestry.

Palaeontologists usually rely on fossils to find out about our ancestors. With modern humans, the study of DNA is a powerful new tool. During development, chromosomes pair up and swap bits of DNA, making each sperm or egg cell unique. This process could make it difficult to trace our genetic history, but it does not occur in mitochondria (tiny energy pumps in our cells, with their own genes) or in Y chromosomes. We inherit mitochondrial genes from our mothers and Y chromosomes from our fathers. Like any genes, they undergo mutations in their DNA at a roughly defined rate. By studying the mutation pattern in people around the world, scientists have worked out a family tree of the world's races. At its base is "Mitochondrial Eve", about 150,000 years ago. She was not the first *Homo sapiens*, but was the most recent common female ancestor we know about. Y chromosomes contain more DNA, so yield a more detailed picture. By studying the pattern of their mutations in the world's indigenous peoples, scientists found "Nuclear Adam", the most recent known common ancestor of all the world's men, who probably lived between 90,000 and 60,000 years ago; they also mapped our spread across the globe.

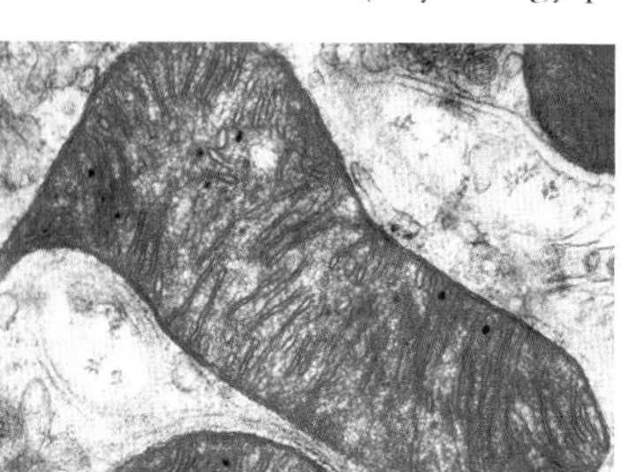

"MITOCHONDRIAL EVE"
By studying the genes in mitochondria, (energy pumps in our cells; one shown above), we think our most recent common female ancestor lived 150,000 years ago.

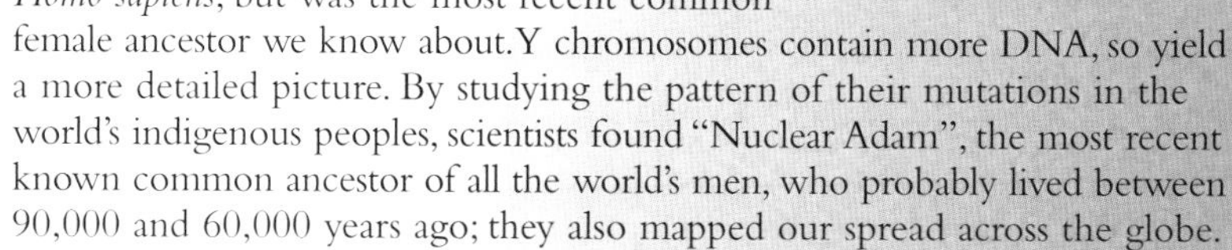

PHYSICAL FEATURES

The early days of *Homo sapiens* are obscure. Very few early modern fossils have been found and there is no clue as to what would trigger the dramatic emergence of culture. Early modern humans had the tall foreheads, domed skulls, and flat faces typical of *Homo sapiens*, and their brains were bigger than those of the majority of their predecessors. Their limb proportions reveal a slender, tropical build unlike that of Neanderthals, so it seems likely they had black skin. Like Neanderthals, however, they would have lived as hunter-gatherers, collecting wild plants, eggs, and honey, and hunting or trapping animals. The hunter-gatherer way of life was to dominate our existence until at least 10,000 years ago, when the Earth's climate became more settled and agriculture was invented. As *Homo sapiens* spread around the globe, the species became increasingly physically diverse. When Cro-Magnons appeared in Europe, for example, they were lighter-skinned than earlier *Homo sapiens* in Africa, probably because they were less exposed to harmful rays of sunlight and their skin therefore needed less protection.

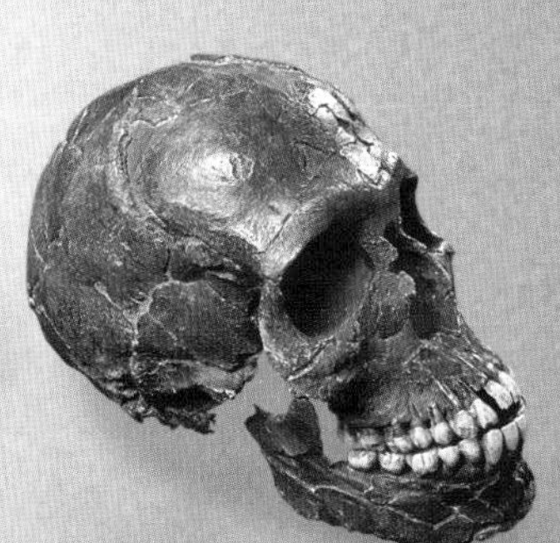

QAFZEH SKULL
Found in the Qafzeh cave in Israel, this 100,000-year-old skull is one of the earliest known examples of Homo sapiens.

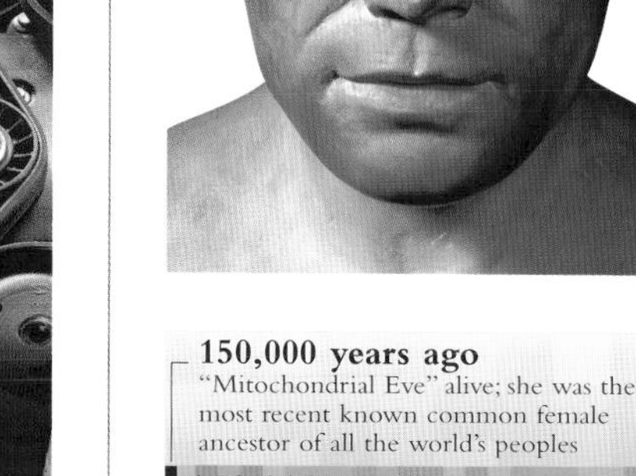

CRO-MAGNON MAN
Cro-Magnon people are likely to have looked physically distinct from early Homo sapiens *in Africa, principally owing to their fairer skin.*

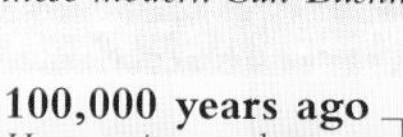

EXODUS FROM AFRICA
When Homo sapiens *first left Africa, they may have looked physically similar to these modern San Bushmen.*

150,000 years ago
"Mitochondrial Eve" alive; she was the most recent known common female ancestor of all the world's peoples

120,000 years ago
Homo sapiens starts to spread across Africa

100,000 years ago
Homo sapiens reaches southern Africa and Israel

150,000 | 125,000 | 100,000

Profile

Charles Darwin

The observations made by naturalist Charles Darwin (1809–1882) while on board HMS *Beagle* during its world voyage led to his ideas on evolution. In 1859, he defined his theory of natural selection in *The Origin of Species*. Darwin was one of the first to suggest that Africa was the homeland of the human race and, when ancient hominid fossils began to turn up, Africa was where they were found. Not a single bone older than 1.8 million years has been found elsewhere.

OUT OF AFRICA

"Mitochondrial Eve", our most recent common female ancestor, is believed to have lived in Africa because the mitochondria of African people have more genetic diversity than those of non-Africans. The results of studies of the Y chromosome also support the idea that *Homo sapiens* originated in Africa. The various non-African populations in the world today seem to have evolved from small subsets of migrating Africans. The oldest modern human bones found come from Herto in Ethiopia and date to 160,000 years ago. By 100,000 years ago, modern humans had evidently spread to southern Africa and Israel (graves dated to that time have been found there), and by 90,000 years ago they were in central Africa, judging from the discovery of intricately carved bone harpoons. As far as can be told, these early people mostly led similar lives to the Neanderthals and used the same limited range of stone tools. According to studies of the Y chromosome, the first wave of migrants began to leave Africa for the farther reaches of the globe around 60,000 years ago, or even a little earlier, spreading along the southern coast of Asia and reaching Australia at least 55,000 years ago. Aboriginal Australians are their descendants, as are the aboriginal peoples of New Guinea, the Andaman Islands, and Sri Lanka. A second wave of African migrants travelled to the steppes of Central Asia perhaps 40,000 years ago, from where they spread into India, Europe, eastern Asia, and Siberia.

GLOBAL SPREAD
Modern humans left Africa in two main waves. The first moved through southern Asia to Australia and the second populated India, the rest of Asia, and Europe. The Americas were one of the last places to be reached.

GENETIC BOTTLENECKS

Compared to other primates, humans have very little genetic diversity. One reason for this is that *Homo sapiens* recently passed through a series of "genetic bottlenecks". A bottleneck occurs when a population falls to a dangerously low level, causing the gene pool to shrink. If the population expands again, all descendants carry copies of the same limited set of genes, preserving a record of the bottleneck for future generations. Drought, famine, and epidemics can all cause genetic bottlenecks. Owing to one or several environmental catastrophes, the number of modern humans may at one point have fallen to as few as 10,000 people. An even more common cause of genetic bottlenecks is migration, and one occurred when modern humans spread out of Africa. In fact, all the world's non-Africans could have descended from a founding population of only 50 people.

GENE POOLS
Africans have the most genetic diversity; Native Americans have a particularly low diversity.

GENETICALLY SUPERIOR?
A single troop of chimpanzees has more variation in certain genes, such as those of mitochondria and Y chromosomes, than do all of the world's people.

90,000 years ago
"Nuclear Adam" in existence sometime between 90,000 and 60,000 years ago; he was the most recent known common ancestor of today's men

75,000

60,000 years ago
Homo sapiens reaches China

55,000 years ago
Homo sapiens reaches Australia

50,000

40,000 years ago
Cro-Magnons appear in Europe

THE GREAT LEAP FORWARD

There are many theories about what triggered the "Great Leap Forward", the jump in creativity that occurred 40,000 years ago in Europe. Because language enables knowledge to be shared, vastly aiding innovation, some think the trigger was a change in the brain or vocal tract that enabled complex language to emerge. Others think it was prompted by the evolution of a sense of self (shown by a belief in the soul and the afterlife), which helped us to know our own feelings and predict those of others – vital to the success of a social primate. A controversial theory is that there was a change in the architecture of the human mind. Some psychologists think that early hominid minds were divided into social, technical, and natural history modules that remained strictly separate. *Homo erectus*, for example, may have had language, but only as part of social intelligence, so could not talk creatively about toolmaking (part of technical intelligence). In *Homo sapiens*, these mental barriers broke down, causing a leap in imagination.

INNOVATIVE LEAP
Cro-Magnons pioneered the use of man-made shelters, such as this one made from mammoth bones.

SEWING IMPLEMENTS
These eyed needles were used some 15,000 years ago to make clothes from animal hides – vital to survive the punishing winters of Ice-Age Europe.

A NEW TOOLKIT

The Cro-Magnons were much more inventive toolmakers than the Neanderthals. They made a wider range of stone tools by striking microliths (thin blades) of flint from a larger core and then adapting these "blanks" for a multitude of different purposes. Cro-Magnons also made innovative use of bone, antler, and wood and, for the first time in history, different materials were combined to make multipart tools. Palaeontologists have found Cro-Magnon fishhooks, net-sinkers, and rope (probably used for snares), all of which reveal the ingenious ways in which they caught and trapped fish and small game. Cro-Magnons are also known to have hunted large animals but, unlike Neanderthals, they killed their prey by throwing spears from a distance rather than fighting them at close range. This saved them from the serious injuries sustained by their Neanderthal cousins.

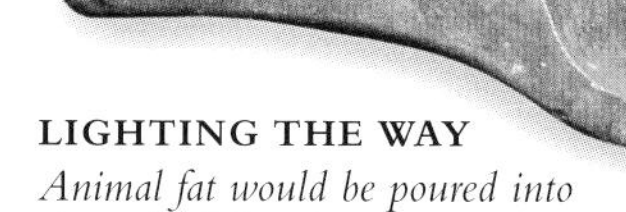

LIGHTING THE WAY
Animal fat would be poured into the well of this Cro-Magnon stone lamp and set alight.

ART AND SYMBOLISM

Even more impressive than the explosion in technology was the appearance of prehistoric art, which reached its zenith 15–30,000 years ago with the famous cave paintings at Chauvet and Lascaux in France, Altamira in Spain, and other caves in the area. The paintings are not primitive; they are strikingly modern and show a fully developed sense of perspective. When Pablo Picasso saw the paintings at Lascaux, he compared the images to modern art, saying "we have discovered nothing". The Cro-Magnons' aesthetic sense is also evident in their love of necklaces, precious stones, musical instruments, and sculptures. Precious stones, such as amber, have been found far from their sites of manufacture – a sign that Cro-Magnon societies traded them with their neighbours. Many of the Cro-Magnons' artistic creations are recognizable as animals or people, but some are more abstract. Among the European cave paintings are geometric patterns of dots and lines similar to Aboriginal Australian rock art, as well as fabulous creatures that are part human and part animal. These abstract images seem rich with symbolic meaning and suggest the existence of an ancient mythology. If the people who created these paintings actually had mythology, then they must also have had articulate language.

MAKING MUSIC
This 25,000-year-old bird-bone flute is an early example of a musical instrument.

LASCAUX PAINTINGS
These world-famous paintings were created around 17,000 years ago in a cave in France.

LADY OF BRASSEMPOUY
This 25,000-year-old masterpiece was exquisitely carved out of mammoth ivory.

DECORATION
The Cro-Magnons created much distinctive jewellery, including this necklace carved from mammoth ivory.

35,000

35,000 years ago
Simple baboon-bone counting device, found in South Africa, dates from this time

30,000 years ago
Homo neanderthalensis disappears

30,000

30,000 years ago
Chauvet cave paintings created by Cro-Magnons, France

28,000 years ago
Sungir bodies buried, Russia

BELIEF IN THE AFTERLIFE

Grave goods reveal that Cro-Magnons must have believed in the afterlife and perhaps had religion. A spectacular example comes from Sungir in Russia: an elderly man and two children were found covered in thousands of ivory beads that were once sewn into clothing. One child had a belt made from 250 fox teeth, an ivory sculpture, a mammoth-tusk lance, and a polished human thighbone packed with red ochre. These people were clearly important and had been buried with grave goods either to protect, entertain, or placate them in the next life – it is hard to know which. Cro-Magnon graves reveal something else: an increased lifespan. Neanderthals were lucky to survive to 40, but Cro-Magnons often lived until well after 60, perhaps reflecting the extra care older people received, and so the extra status they may have had.

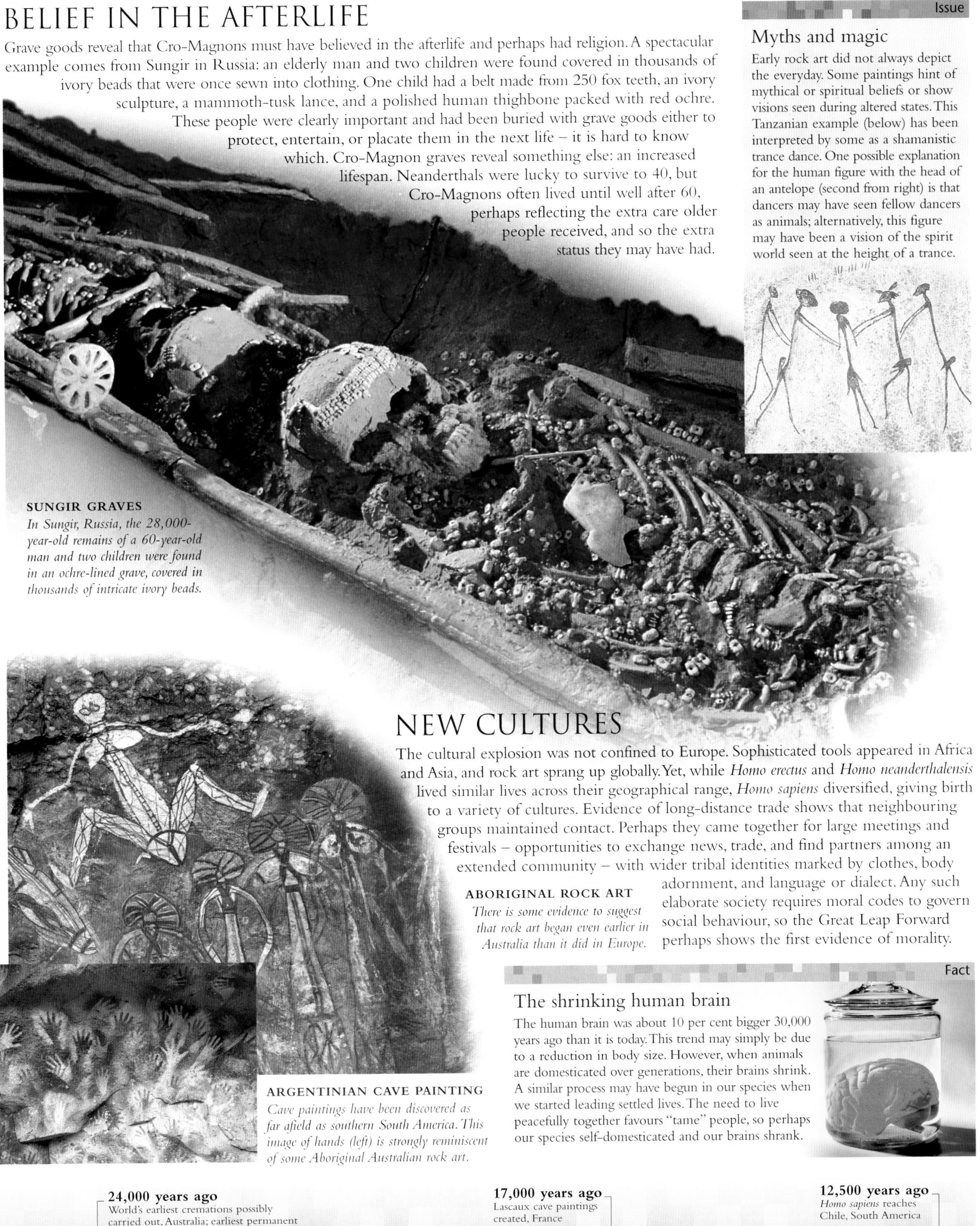

SUNGIR GRAVES
In Sungir, Russia, the 28,000-year-old remains of a 60-year-old man and two children were found in an ochre-lined grave, covered in thousands of intricate ivory beads.

Issue

Myths and magic

Early rock art did not always depict the everyday. Some paintings hint of mythical or spiritual beliefs or show visions seen during altered states. This Tanzanian example (below) has been interpreted by some as a shamanistic trance dance. One possible explanation for the human figure with the head of an antelope (second from right) is that dancers may have seen fellow dancers as animals; alternatively, this figure may have been a vision of the spirit world seen at the height of a trance.

NEW CULTURES

The cultural explosion was not confined to Europe. Sophisticated tools appeared in Africa and Asia, and rock art sprang up globally. Yet, while *Homo erectus* and *Homo neanderthalensis* lived similar lives across their geographical range, *Homo sapiens* diversified, giving birth to a variety of cultures. Evidence of long-distance trade shows that neighbouring groups maintained contact. Perhaps they came together for large meetings and festivals – opportunities to exchange news, trade, and find partners among an extended community – with wider tribal identities marked by clothes, body adornment, and language or dialect. Any such elaborate society requires moral codes to govern social behaviour, so the Great Leap Forward perhaps shows the first evidence of morality.

ABORIGINAL ROCK ART
There is some evidence to suggest that rock art began even earlier in Australia than it did in Europe.

ARGENTINIAN CAVE PAINTING
Cave paintings have been discovered as far afield as southern South America. This image of hands (left) is strongly reminiscent of some Aboriginal Australian rock art.

Fact

The shrinking human brain

The human brain was about 10 per cent bigger 30,000 years ago than it is today. This trend may simply be due to a reduction in body size. However, when animals are domesticated over generations, their brains shrink. A similar process may have begun in our species when we started leading settled lives. The need to live peacefully together favours "tame" people, so perhaps our species self-domesticated and our brains shrank.

25,000

24,000 years ago
World's earliest cremations possibly carried out, Australia; earliest permanent huts built with clay roofs, Europe

23,000 years ago
Lady of Brassempouy sculpted, France

20,000

17,000 years ago
Lascaux cave paintings created, France

16,000 years ago
First huts built with mammoth-bone roofs, western Russia

15,000

13,500 years ago
Homo sapiens crosses into the American continent

12,500 years ago
Homo sapiens reaches Chile, South America

SETTLED SOCIETIES

Between 10,000 and 1000BC, humanity began to leave a bigger impression on the natural world. The changing climate and environment at the end of the last ice age offered new possibilities for human life. Agriculture emerged and was sustained by a new type of cereal and irrigation technology. Agricultural surplus gave rise to settled agrarian communities, then to cities with a wider division of labour. The need to store this surplus against bad times produced pottery and recording systems, and there is also evidence of elaborate burial rituals. Humanity did all these things in river valleys across the world, and even when specific cultures were shortlived, the struggle to sustain a settled life was uninterrupted.

SETTING BOUNDARIES
This Babylonian boundary stone constitutes an early legal document. Written evidence of this kind allows us to read the story of our past.

THE WRITTEN WORD

Without written records, there would be no human story or "history". Written symbols are sophisticated tools that allowed humanity to refine skills, ideas, and understanding over thousands of generations. The first evidence of this capacity to record things in writing comes in the form of Sumerian clay counting tokens and written symbols (cuneiform) from around 3400BC. The wide variety of types of symbol indicated they were used across a range of activities, from accounting to recording creation stories and cosmologies, and writing soon became a necessity in societies that stored and traded goods. Early symbols were simple pictorial representations of common sources of livelihood, such as a horned head for an ox; more abstract systems, such as hieroglyphics and early alphabets, developed to express more complex ideas. Those who possessed the jealously guarded ability to use and interpret written language were able to establish the law.

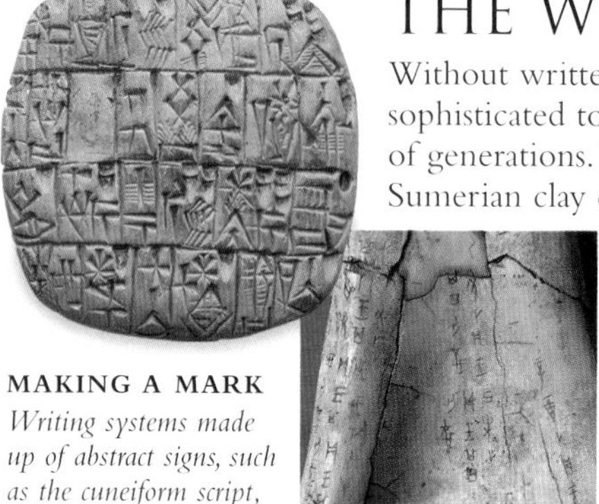

MAKING A MARK
Writing systems made up of abstract signs, such as the cuneiform script, enabled the expression of complex ideas.

ORACLE BONES
Questions about the future engraved on bone are the earliest known form of Chinese writing.

EARLY AGRICULTURE

Around 12,000 years ago, the cultivation of plants for food began in the Fertile Crescent in Mesopotamia (modern-day Iraq) when humans witnessed and then repeated accidents of nature. A chance crossing of wild grasses had produced a new hybrid strain with a much fuller head of seed. Scattered by the wind, the hybrid eventually crossed with another grass to create "bread wheat", an even richer grain ideal for human cultivation. As a result, agriculture sprang up in fertile regions: near lakes, rivers, the sea, or in areas with sufficient rainfall at the crucial times of year. However, as farming methods spread and larger populations became dependent on crops, people were forced to attempt an agricultural lifestyle in less fertile areas. Irrigation techniques were vitally needed in arid, unpredictable climates. The methods were basic at first – a case of digging a well and carrying water to the crops – but they soon became increasingly sophisticated. Networks of canals were built to store water from seasonal downpours and to channel it to crops, or even to divert water from lakes and rivers farther afield. Similar methods were used to drain waterlogged regions.

SEEDS OF CHANGE
A new hybrid bread wheat required human cultivation. Tools such as sickles (left) and hand mills (far left) made the task easier.

DOMESTICATION
The domestication of plants and animals formed the basis of farming, which was transformed by inventions like the plough. This Egyptian fresco dates to c1300BC.

THE WHEEL
The invention of the wheel transformed agriculture, and also trade and military conquest; this example dates from c3500BC.

Fact

The significance of pottery

Earthenware storage vessels are a classic feature of early Asian, Middle Eastern, and American civilizations. Within a few hundred years of the advent of agriculture, pots emerged to store the surplus food produced. The earliest vessels were found not in the Fertile Crescent, but in Japan's Honshu district. Similarly dated pots found in river valleys all over the world show that civilizations developed at the same time around the globe.

EARLY JAPANESE POT

10,000BC First known use of pottery vessels, Honshu, Japan

9000BC Settled agriculture established; the first selected grasses produce an early form of wheat

8000BC Sheep and goats domesticated in the Middle East and pigs in China

7500BC The world's first walled town established, Jericho, Israel

6500BC Cattle successfully domesticated in Sahara region, North Africa

6000BC Corn (maize) first cultivated, Ecuador

10,000BC | 9000BC | 8000BC | 7000BC | 6000BC

TOWER OF JERICHO
To give a vantage point, a community, such as that of Jericho, would construct a watchtower at the highest point of the wall surrounding its settlement.

THE FIRST SETTLEMENTS

The possibility of growing food almost anywhere, and so living almost anywhere, transformed human society. In fact, the advent of agriculture led to the simultaneous rise of civilizations all over the world. In the past, staying in one place had probably meant death; now it tended to mean security. Jericho, a farming settlement just north of the Dead Sea, seems to have been the first walled human community, dating from around 7500BC. Initially, such settlements were agricultural, comprising the homes of those who farmed the land in the immediate vicinity and came together for protection. These communities could expand as it became technically possible to build bigger settlements, and because larger communities could grow and store a greater surplus of food, they were able to expand even further. After a certain level of growth, some community members could be freed to undertake more specialized roles, such as those of priests, potters, or soldiers. Hierarchies began to form, with the interpretation of knowledge and physical protection of a settlement becoming privileged roles that were respected by those who produced and traded food.

CATAL HUYUK
In early cities, such as this Turkish settlement from around 7000BC, many people could live together carrying out complementary tasks.

A NEW SOCIAL ORDER
As societies formed, some people gained status and had to protect their position, even after death. This model army guards the tomb of an Egyptian ruler.

BURIAL

The high status of certain individuals in these new societies is reflected in the manners of their burial. Across Asia, the "elite", who may have been members of the military caste or seen as gods or gods' messengers on Earth, were buried under huge mounds; in Egypt and areas of Mesoamerica, leaders were laid to rest within elaborate pyramids or mausoleums. In some areas, most bodies were mummified for preservation; in others, they were simply buried, cremated, or left exposed for birds to pick at the bones. Bodies of the social elites were invariably surrounded by a range of grave goods, which sometimes even included sacrificed servants, concubines, soldiers, or animals – although over time these were replaced with artificial representations, such as model armies. Choice of grave goods can tell us much about the religious beliefs of a society, the position a deceased person held, and which materials were considered valuable at that time. Burial rituals were intended to ease the passage of the dead to the afterlife; to celebrate their wealth, distinction, and achievements; and to give a sense of continuity to those who remained.

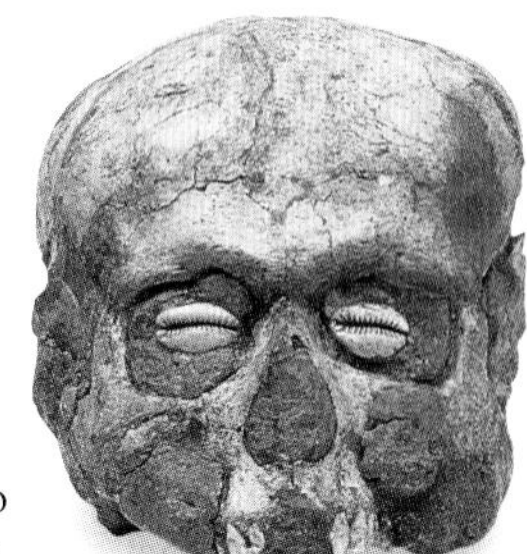

BURIAL RITUALS
Shells in the eye sockets of this Jericho woman's skull indicate that a burial ritual may have been performed.

CREMATION URN
The Harappan civilization from the Indus Valley (3000–1500BC) cremated bodies and stored the ashes in funerary urns.

Fact

Building the pyramids

According to the Greek historian Herodotus, it took 400,000 men a total of 20 years to construct the Great Pyramid at Giza in Egypt. This awesome human achievement stands 147m (482ft) tall and comprises some 2.5 million blocks of stone, each weighing an average of 2.5 tonnes (2.8 tons). The blocks were manually rolled up earth ramps to form "stepped" layers. Finally, the step effect was hidden by applying a smooth layer of limestone.

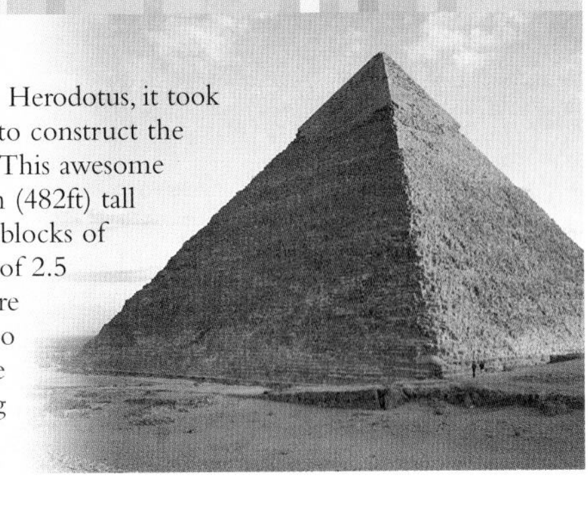

5500BC – First irrigation system developed, Mesopotamia
5000BC – First use of copper, Mesopotamia
4500BC – First large cemeteries, Europe
4000BC
3400BC – Use of clay counting tokens and first written symbols, Sumeria
3200BC – First wheeled transportation, Asia
3000BC – Silk first produced, China
2750BC – Start of great period of pyramid building, Egypt
2000BC
1500BC – First evidence of metal-working, Peru
1360BC – Assyrian empire begins to build up, western Asia

THE CLASSICAL WORLD

From 1000BC to AD400, civilizations remained susceptible to decay and attack, although most nomadic invaders formed new, settled populations. As societies achieved greater levels of prosperity, they found new scope for reflection and, in some places, science and technology emerged. The greatest development was that of a political or civil society, with the emergence of concepts such as citizenship and democracy. Cultural evidence, from Homer's poetry to the Olympics, shows how people understood and celebrated their world. The forms of religion and philosophy we are familiar with also arose at this time: Buddhism and Confucianism were established and Jewish monotheism gained a new form with the birth of Christianity.

THE GREAT WALL
China's Han empire extended this already-existing defensive wall to guard itself from attack.

AN AGE OF EMPIRES

At the start of the first millennium BC, the Assyrian empire, covering western Asia, was probably the world's greatest. It was destroyed by the Babylonians around 606BC, who were in turn conquered by the expanding Persian empire in 539BC. Persia dominated the world until about 325BC, when Alexander the Great conquered lands from Greece to the borders of India. Yet within 100 years of his death in 323BC, his empire had divided and disappeared. By the beginning of the 2nd century AD, four great empires reigned. Rome ruled the Mediterranean, parts of northern Europe, the Middle East, and Africa's north coast. In the east was the Han Chinese empire. In between, the Parthian empire covered much of western Asia and the Middle East, while the Kushan stretched from the Aral Sea through present-day Uzbekistan, Afghanistan, and Pakistan, and into northern India. By AD400, the demise of Rome and Han China had led to a power vacuum. Much of the world entered a state of chaos, with the recurring destruction of cities by barbarian attack.

ALEXANDER THE GREAT
Alexander, King of Macedonia in northern Greece, rose to power in 330BC. In little more than 10 years, his armies had conquered much of the world known to him.

ECONOMIC EXPANSION

Coinage was invented in the 7th century BC in Greek Asia Minor as a convenient method of payment. India and China saw similar developments from 500BC and, by 200BC, coins had spread to western Europe. Coins are not only important for local payment: long-distance trade is more efficient if conducted through a medium of exchange rather than by the direct trade of goods, as long as the medium is portable and of mutually acknowledged value. The world's empires were linked by trade routes that skills and goods travelled along. From Spain in the west, trade routes skirted the Red Sea and went down the African coast. Others went from Africa to India, crisscrossed the Mediterranean and the Middle East, and went across the Asian steppes to China. Some even extended to the Pacific. Trade and military power were often linked, and wars were fought over raw materials and strategic routes.

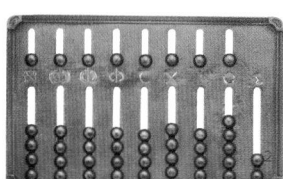

ROMAN HAND ABACUS
This method of calculating complex payment facilitated trade between individual community producers and also between empires.

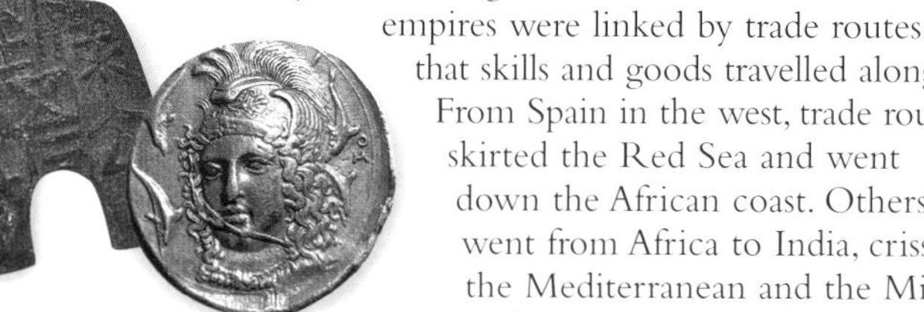

SIGNS OF WEALTH
Early coins, such as these Chinese and Greek ones, were made from precious alloys and often carried a ruler's image to mark their value.

EARLY ROMAN SHIP
Trade benefited from better ships and navigational aids, which were in turn the result of increased wealth.

CLASSICAL CULTURE
The "golden age" of Greek classical culture is celebrated in Raphael's painting "School of Athens", which features Aristotle, Plato, Socrates, and Ptolemy, amongst others.

1000BC | 900BC | 800BC | 700BC | 600BC | 500BC | 400BC

- **850BC** Homer believed to have created his *Iliad* and his *Odyssey*, Greece
- **776BC** First Olympic Games, Greece
- **753BC** Rome founded
- **650BC** Coinage invented in Greek Asia Minor
- **606BC** Assyrian empire destroyed by Babylonians
- **560BC** Buddha born, Nepal
- **539BC** Babylonian empire conquered by Persians
- **490BC** Battle of Marathon during the Persian–Greek wars
- **432BC** Parthenon completed, Athens
- **334BC** Alexander the Great, controller of Greece, invades Asia Minor

EMERGENCE OF POLITICAL SOCIETY

The idea of a citizen's rights, responsibilities, freedoms, and duties developed in the city state of classical Athens under the stewardship of Pericles (about 495–429BC). Greek political thinkers, such as Plato (427–347BC) and Aristotle (384–322BC), went on to define not only how society was, but also how it ought to be. In *The Republic*, Plato set out his ideas for a utopian society. In *Politics*, Aristotle took a more scientific approach, comparing existing political states so as to formulate arguments for the best balance in the administration of public life. Together, they created and disseminated the rich language of modern politics, including concepts such as the citizen, the assembly, the republic, democracy, oligarchy, tyranny, and dictatorship. The vocabulary and framework of modern philosophy were similarly set out. In fact, the Greeks made no real distinction between political and moral philosophy, determining that political action and institutions and ethical conduct were all integral to life. At the same time in China, Confucius and his followers were also searching to understand the relationship between the individual and society.

EASTERN PHILOSOPHY
Confucius (551–479BC) developed an influential view of how humanity should relate to the world, which was centred on duty to the state.

PERICLES
From the relatively short rule of this statesman, we derive many of our ideas about politics and public life.

THE AGORA IN ATHENS
At the heart of ancient Athens, the Agora (or "market") was the centre of the state justice system, political and commercial life, administration, and religious activity.

HUMAN FOCUS
Greek sculpture took the individual as an object of study, focusing on and celebrating the human form.

KNOWLEDGE AND CULTURE

A cultural revival around 800BC accompanied the emergence of city states across Greece. Greek culture gradually spread far beyond the boundaries of ancient Greece, influencing cultures all around the globe. Worldwide, scientific and technological developments comprised both abstract systems of knowledge and practical tools to facilitate life, agriculture, and trade – from Pythagoras' theorem, which defined the relationship of objects in space, to Archimedes' irrigation device and Chinese paper-making techniques. The desire for order in this period was also evident in an astronomy that sought not only harmony in the stars, but to provide for seafarers and travellers. Perhaps even more important for human history was the emergence of a new concern for humans themselves, a feature of classical Greece that Renaissance scholars would later seize upon. Although ancient civilizations intimately linked individual human life to a cosmic order, fate, destiny, and the will of the gods, the classical world came to believe that society was directed by humans. In-depth consideration of the individual and the consequences of his or her actions is at the very heart of what we understand as "modern".

PRACTICAL INNOVATION
The remains of Roman underfloor heating systems (above) and those of piping systems show how advanced classical society was.

Profile

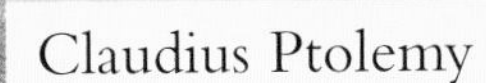

Claudius Ptolemy

The astronomer, mathematician, and geographer Ptolemy (AD87–150) lived in Alexandria, Egypt, a principal centre of Greek classical culture. Ptolemy rationalized the order and apparent movements of the planets known at the time. His elaborate patterns of astronomy were based on Aristotle's belief that the Earth was fixed, with the Sun, stars, and planets revolving around it. The Ptolemaic universe is now often derided because it works on the supposition that the Sun revolves around the Earth. However, it is incredible to think such a sophisticated idea came so early.

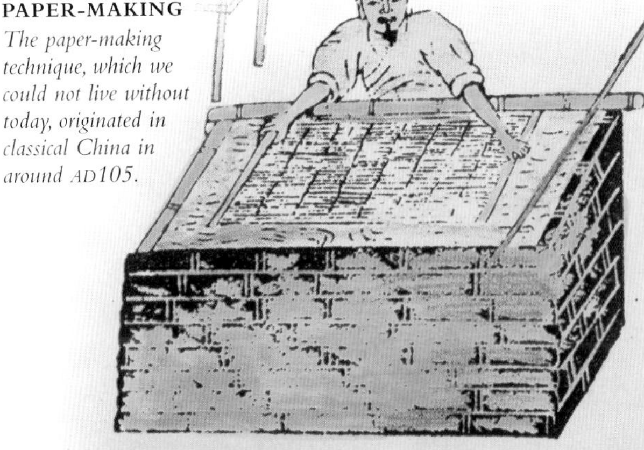

PAPER-MAKING
The paper-making technique, which we could not live without today, originated in classical China in around AD105.

300BC | 200BC | 100BC | 0 | AD100 | AD200 | AD300

247BC Parthian empire emerges

221BC Shih Huang-ti of the Ch'in dynasty unites China

146BC Rome conquers Greece and Carthage, becoming master of western Mediterranean

30BC Antony and Cleopatra die; Egypt annexed to Roman empire

5BC Jesus is born

AD43 Start of Roman conquest of Britain

AD79 Eruption of Vesuvius destroys the cities of Herculaneum and Pompeii

AD105 Paper is invented in Han China

AD228 End of Parthian empire

AD230 End of Han dynasty in China: China first divides into three states, then fragments

AD300 Abacus in use, China

AD330 Constantinople (Byzantium) becomes capital of Roman empire

ORIGINS

WORLD AT A CROSSROADS

The period between the fall of Rome and the Renaissance is often termed the "Middle Ages". After a time of uncertainty and cultural loss, a period of creativity set in motion long-lasting developments based on the achievements of the classical world. This era saw the last great nomadic incursion against civilization, that of the Mongols, and the bubonic plague. There was a rapid expansion of Islamic culture, which cut the Chinese off from their western trade routes and forced Europe to turn west to reestablish trading links with the East. It was also through Islamic culture that the science, philosophy, and investigative outlook of the ancient world were renewed, developed, and dispersed, providing the foundations for the Renaissance.

UNIVERSITY LIFE
Universities were founded in the new European centres of culture; this manuscript image shows a lecturer at Bologna University, one of the first.

THE SPREAD OF ISLAM

The scale and pace of Islamic expansion during the Middle Ages were unprecedented. When the prophet Muhammad died in 632, his authority extended only around Mecca and Medina, and perhaps to central and southern Arabia. Over the next century, Arab armies carried Islam as far west as Spain, and as far east as northern India and the Chinese border. The Arab armies emulated and surpassed Alexander the Great's achievements, and their influence lasted much longer. Their religious conviction swept aside the crumbling Byzantium (late Roman) and Persian empires. Yet the Arab expansion was not merely a military adventure underpinned by religious enthusiasm. It helped to sustain civilization, with learning and culture supported at a far more advanced level than in any other contemporary civilization, even that of the monasteries of Europe. Although the Islamic world soon lost its unity, it had an enormous influence on modern culture.

A PROPHET'S BIRTH
This illustration depicts the birth of the prophet Muhammad in about 570. The religion he expounded was to have enormous global influence.

DIFFUSION OF CULTURE
The extent of Islam's spread is evident in southern European Islamic architecture, such as the Alhambra, Spain. Engineering, science, philosophy, and art flourished under Islam.

THE MONGOL INCURSION

Emerging from the depths of Asia, the Mongols were the last pastoralist and nomadic people to make a major and successful incursion against the settled civilizations of the world. In 1206, Genghis Khan (then a tribal chief) succeeded in uniting the Mongol tribes, gaining control of an army powerful enough to destroy the Ch'in empire in China. His successors went on to recast the map of the world. They advanced into Russia, swept through Persia's central Asian territories, inflicted the first significant defeats on the centres of Islam, and later invaded India. In 1241, Christian Europe was saved from invasion only by the death of the Great Khan, Maijke. Mongol power lasted until about 1405, but its social and cultural legacy is nothing compared with the achievements of the Arab diaspora. Despite their enormous military impact, Mongol influence survives only through their assimilation into the settled societies they conquered – in a dynasty of Chinese rulers and in the Moghul empire in the Indian subcontinent.

GENGHIS KHAN
The Mongols' greatest leader was born around 1162. His death in 1227 did not halt the Mongol advance.

BRIDGING WORLDS
The period between the end of the Roman empire and the Renaissance was a bridge between the ancient and the modern, in which classical achievements were built on.

400 | 500 | 600 | 700 | 800 | 900

- **410** Visigoths sack Rome, leading to collapse of Roman empire
- **404** Latin version of the Bible completed
- **600** Windmills in use, Persia
- **650** Koran written, 18 years after death of the prophet Muhammad
- **661** Start of first Muslim dynasty, the Umayyads
- **730** Paper-making spreads from China to Muslim world
- **750** Gunpowder invented, China
- **760** Arabs develop algebra and trigonometry
- **800** Printing with blocks developed, Japan
- **850** Collapse of classic Mayan culture in Mesoamerica

THE IMPACT OF PRINTING TECHNOLOGY

The precursor to modern printing was xylography, the art of printing from carved wooden blocks. This process originated in China and began to appear in Europe during the last quarter of the 14th century. Printing technology developed very rapidly in 15th-century Europe. The advent of movable metal type was a revolutionary advance that marked the start of mass communications because it allowed texts to be set in relief much more quickly than when carved wooden blocks were used. The invention of typography, based on the use of movable type, is credited to Johannes Gutenberg in approximately 1450, and he produced the first printed bible approximately 6 years later. Mechanical printing was the gateway for an unprecedented diffusion of knowledge. Access to information was not only wider in terms of a greater quantity of printed materials, but also in the sense that books and pamphlets were now frequently presented in vernacular language. As the technology spread, so the price of books fell, with the result that more and more people learned to read. In a very real sense, the world expanded – or became more intelligible – for a larger number of people.

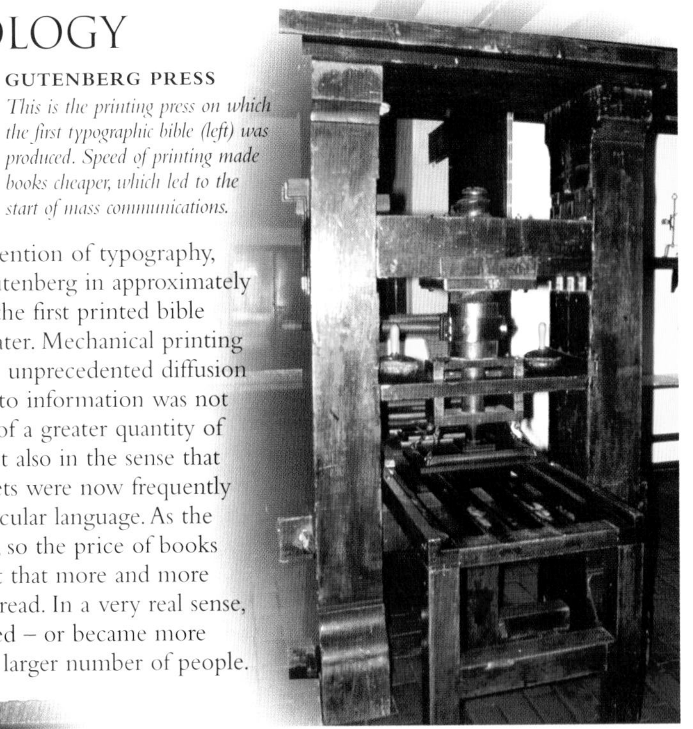

GUTENBERG PRESS
This is the printing press on which the first typographic bible (left) was produced. Speed of printing made books cheaper, which led to the start of mass communications.

EASTERN PRINTING
This early Japanese wooden printing block shows that printing technology was developed the world over.

FIRST PRINTED BIBLE
The Gutenberg Bible was printed on the first typographic printing press (right) around 1456. From this time onwards, there was an avalanche of printed material.

GUNPOWDER "REVOLUTION"

The fierceness and stamina of the Mongol army had inspired shock and awe, but such primitive energy ceased to be a threat to better-established societies armed with gunpowder and firearms. Discovered in China, the formula for gunpowder was used to produce explosives at least from the time of the Sung dynasty (960–1279). The formula had clearly reached Europe by the second half of the 13th century because it is referred to several times in the work of the English scientist Roger Bacon. The first recorded evidence of using gunpowder in handguns is at the Italian town of Forli in 1284, and there are frequent references to siege guns employing the new form of explosive from the early 14th century. Such early devices did not make for much of a revolution in military tactics and strategy but, within a century, guns became much more effective. In fact, the technological superiority of the European armies was a decisive factor in the establishment of European domination and colonial power in the Americas.

EXPLOSIVE ALCHEMY
A 16th-century woodcut shows German monk and alchemist Berthold Scharz at work producing gunpowder in the early 14th century.

Fact

The Black Death

Carried by rats from China, bubonic plague caused the death of between a third and a half of Europe's population in the last half of the 14th century. It is likely to have arrived through a military incursion on Europe or on merchant ships. Surprisingly, the massive loss of life probably contributed to the rapid development of European domination in the world. As land became more abundant and labour scarcer, old feudal hierarchies became difficult to sustain, making cities more dynamic. It was easier for survivors to gain an economic surplus, and society in general became more at ease with change.

EARLY GUNS
This type of cannon revolutionized military tactics once Europe had acquired the gunpowder formula.

- **984** Single or "flash" lock-gates used on canals, China
- **1000** Camera obscura invented, Arabia; expansion of Inca empire, Peru; Vikings reach America
- **1080** Magnetic compass invented, China
- **1126** Arabian philosopher Averroes born
- **1150** Completion of temple of Angkor Wat, Cambodia; paper-making reaches Europe
- **1175** First Muslim empire in India
- **1206** The Mongols under Genghis Khan begin their conquest of Asia
- **1271** Marco Polo travels the Silk Road between Europe and China (until 1295)
- **1341** The Black Death begins in Asia; goes on to ravage Europe in 1348
- **1453** Ottoman Turks capture Constantinople and bring Byzantium empire to an end
- **1492** Muslim rule ends in Spain

NEW HORIZONS

The period known as the Renaissance, which had begun around 1400, was a journey of discovery and debate that pushed back all known limits. From the late 15th century, scholars were looking at the world very differently, believing that humanity determined its own fate. Experimentation and observation became essential in all fields: artists observed their subjects and developed a sense of perspective; seafarers watched the stars and mapped the Earth; engineers solved life's practical problems; philosophers developed a new world order. Italy's city states played a critical role in this cultural revolution, but by the late 1500s, as European attention moved towards transatlantic expansion, Spain and France were leading powers; Britain was waiting in the wings.

AHEAD OF ITS TIME
Brunelleschi's dome (built 1419–1436) in Florence, is a miracle of engineering and artistic imagination that set the tone for the Renaissance.

SCIENTIFIC ADVANCEMENT

Like the Greeks, Renaissance scientists delighted in knowledge, but they sought to apply it more systematically. Arabic science provided the tools and methods to understand the world, and mathematical measurement and calculation became the basis of all science. Scientists moulded a relationship between mathematics, astronomy, and navigation, and used their knowledge of the planets and mapping methods to chart new lands. Wealth from this "new world" helped the scientific revolution to spread more widely. Astronomical research was severely restricted, however, because the Catholic Church declared it heretical to question the Earth's place at the centre of the universe. Yet both Copernicus (1473–1543) and Galileo did just that, showing that the Earth was spherical and in uniform motion around the Sun. They removed science from the sphere of Church authority.

COPERNICAN WORLD SYSTEM
Copernicus revolutionized astronomy, showing that the Earth rotated daily and moved around the Sun once a year.

GALILEO'S TELESCOPES
Observation of the planets led Galileo (1564–1642) to argue for separating theology and science.

EXPLORATION AND DISCOVERY

The voyages of discovery for which the Renaissance is renowned were inspired more by the pursuit of trade and wealth than knowledge or adventure. Turkish domination of the Middle East had cut off European access to East Indian spices, so European monarchs hired navigators to find new routes to reestablish this trade. These voyagers discovered lands that fired European greed and imagination. Initially the voyages were tentative because sailors had trained in the Mediterranean where they were never far from a landfall: early navigators such as Vasco da Gama set out south for the Indies, skirting Africa and the Cape of Good Hope. Although this route proved valuable, more intrepid crews (including those of Christopher Columbus and Ferdinand Magellan) steered ships west. Sailors navigated using the stars and a compass, and charted their passage on a quadrant so they could follow it again. Such devices were the raw material of map-making. Europe's future lay to the west, but its exploration had a detrimental impact on the native populations of the "New World".

COLUMBUS
Italian sailor Christopher Columbus (1451–1506) believed he could reach China by sailing across the Atlantic Ocean. He persuaded the rulers of Spain to back him.

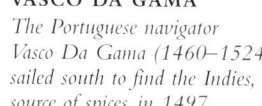

VASCO DA GAMA
The Portuguese navigator Vasco Da Gama (1460–1524) sailed south to find the Indies, source of spices, in 1497.

PILGRIM FATHERS
After the discovery of new territory, many Europeans moved to colonize it. This image shows passengers travelling to America on board the Mayflower.

Fact

The Golden Triangle

The transatlantic slave trade, known as the Golden Triangle, began in 1441 when a Portuguese sailor seized 10 Africans to take to the Americas. In 1454, Pope Nicolas V officially started the trade. Merchant ships would leave Europe for West Africa laden with manufactured goods. In West Africa, these were exchanged for slaves from the interior who were transported to the Americas. Those who survived the journey were then sold to plantation owners in exchange for valuable produce, such as cotton, tobacco, molasses, and rum.

1493 Columbus reaches America

1497 Vasco da Gama sets off for the Indies from Lisbon

1500

1519 Cortés arrives in Mexico; Aztec empire collapses

1520 Chocolate introduced to Spain from the Americas

1520

1528 Mombasa revolts against Portuguese rule

1530 Coal mining begins in Europe

1540

1545 Discovery of silver at Potosí, Bolivia

1550 Beijing is besieged by the Mongols for a week

1560

1560 Reunification begins in Japan, a land of warring nobles

1570 Potato introduced into Europe from the Americas

1580

1588 Famine and pestilence sweep China

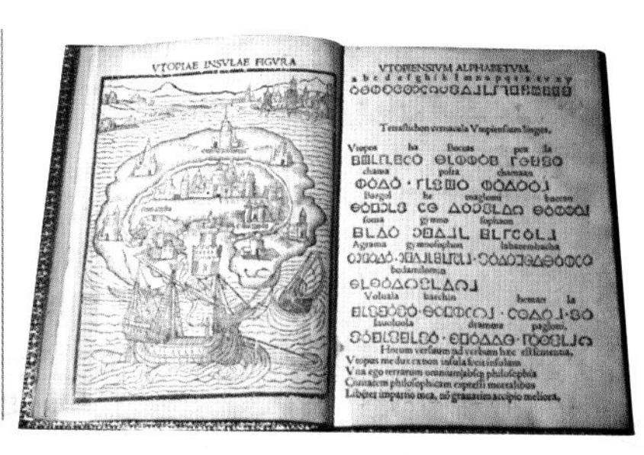

THE FIRST "UTOPIA"
Thomas More coined the term "utopia", in his 1516 book, to refer to a desirable yet unattainable place. The book linked the voyages of discovery to journeys of the mind.

A CHANGING PHILOSOPHY

Renaissance thinkers, scientists, and artists went on an even greater voyage of discovery than that of the explorers – to learn about mankind and push back the limits of humanity. A new feeling of self-determinism arose, with philosophers stating that humans were not simply pawns in a preordained cosmic order. This humanist stance became increasingly widespread and led to a new urgency and quality of artistic production, and to increased interest in scientific and geographical exploration. To the Greeks and Romans, chance had referred to the will of the gods. The Renaissance concept of chance, in contrast, held that a person made his or her own destiny by seizing opportunities and taking decisive action, which brought about a new sense of optimism and hope for the future. It is difficult to date this shift in perspective but, in the space of 350 years between 1350 and 1700, an enormous transformation in social outlook and philosophy did come about. Geographical exploration had opened up new vistas in the physical world and corresponding horizons in the human mind. The presence of God was still important and remained beyond the limits of human understanding, but religion was increasingly held open to enquiry and analysis.

DESIDERIUS ERASMUS
Contemporary and confidant of Thomas More, Erasmus (c1469–1536) is considered prince of the humanists. He expounded the view that man was his own maker.

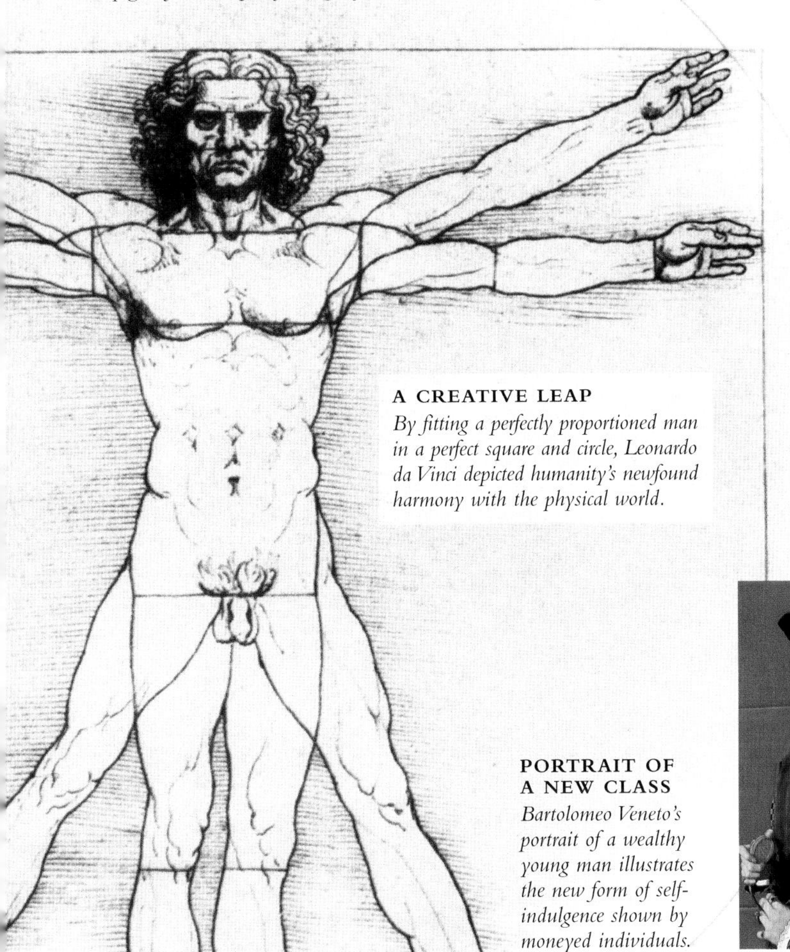

A CREATIVE LEAP
By fitting a perfectly proportioned man in a perfect square and circle, Leonardo da Vinci depicted humanity's newfound harmony with the physical world.

THE ARTS

Although the social order was still prescribed and social mobility remained extremely limited, advances in trade and economic production, the growth of towns, and the decay of feudal hierarchies allowed a new concern for the individual to surface. A new urban elite developed, and these wealthy people not only prospered but reflected upon themselves and were proud of their own achievements. Therefore, as patrons of the arts and celebrators of their own success, members of the new elite were just as likely to commission a portrait of themselves as yet another one of the Virgin Mary. Such elites also owned their own books and, although they were certainly proud of their classical learning, these books were frequently written and published in common, everyday language rather than in Latin. Thanks in part to the earlier influences of Dante, Chaucer, and the printing press, and now – especially in England – to Shakespeare, the rich variety of human experience began to be communicated in the concentrated forms of language that we continue to use today all around the world. At the same time, theatre-going became an outlet for the hopes and fears of generations of Europeans from a wide variety of socially diverse backgrounds.

PORTRAIT OF A NEW CLASS
Bartolomeo Veneto's portrait of a wealthy young man illustrates the new form of self-indulgence shown by moneyed individuals.

Profile

William Shakespeare

The playwright Shakespeare (1564–1616) was born in Stratford-upon-Avon, England. He wrote his plays, such as *Hamlet* and *Romeo and Juliet*, during the Elizabethan age, when English society was unusually dynamic. Trade, discovery, and innovation had transformed a static medieval world and, in Shakespeare's work, there is a sense both of a vanished past and the exhilaration of discovery. The form and content of his drama reached out to and touched a socially disparate audience as never before. Shakespeare possessed a wide vocabulary and even invented words now in common usage in the English language.

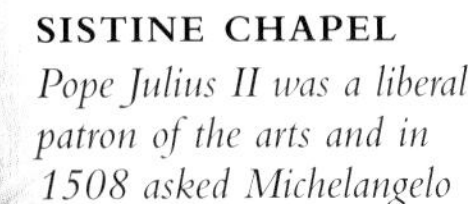

SISTINE CHAPEL
Pope Julius II was a liberal patron of the arts and in 1508 asked Michelangelo to paint the ceiling of the Vatican's Sistine Chapel.

1600 | 1620 | 1640 | 1660 | 1680

- **1600** British East India Company founded
- **1602** Dutch East India Company founded
- **1608** Telescope invented by Hans Lippershey, Holland
- **1630** Circulation of blood discovered by English physician William Harvey
- **1644** Qing dynasty, last of Chinese imperial dynasties, established
- **1648** Construction of Taj Mahal completed after 16 years
- **1655** Pendulum clock invented by Christian Huygens, Holland
- **1660** Gujaratis make earliest known Indian nautical charts
- **1682** English astronomer Edmund Halley observes Halley's comet
- **1688** William Dampier is the first Englishman to visit Australia

A CHANGING WORLD

Agricultural and industrial revolutions brought rapid developments in human technical capacity, productivity, and the mass availability of goods, yet such technological advances were outstripped by developments in the political and social imagination. By 1815, the UK (formed 1801) had taken centre stage in the world scene, and optimism and a sense of progress prevailed in Europe. Within 100 years, Germany and the US had challenged British hegemony, and the world had changed immeasurably. Millions died in two world wars, empires collapsed, and the new power centres (the US and USSR) lived in ideological conflict. Political turmoil, economic instability, and human devastation continued through the 20th century, and there is, today, a renewed sense of unease.

Fact

The Lisbon earthquake

On 1 November 1755, the populous city of Lisbon was destroyed by an earthquake. The event stimulated philosophical debate about God, fate, and human intervention. Jesuits saw it as God's retribution for human sins; others, as a powerful argument against His existence. Some said that people could avoid such disasters by moving from large cities. Other consequences were also significant: the Portuguese dictator rejected the Jesuits' argument and cut them off from influence, a precursor to the rapid secularization of European governments after the French Revolution.

ENLIGHTENMENT AND UNCERTAINTY

ENCYCLOPÉDIE,
OU
DICTIONNAIRE RAISONNÉ
DES SCIENCES,
DES ARTS ET DES MÉTIERS.

DIDEROT'S ENCYCLOPEDIE
French philosopher Denis Diderot compiled this vast record of the Enlightenment between 1745 and 1772. It aimed to represent all knowledge and, through this, combat fear.

The 18th-century "Enlightenment" was founded on the belief that by understanding life, the world, and its social institutions, the human condition could be improved. Until this era, people had either looked back to the Golden Ages of classical Greece and Rome or settled for a better life in the next world. Now, greater material prosperity and an escape from the worst ravages of disease brought new optimism. This quest for knowledge and scientific order continued into the 19th century, but the findings began to unsettle society. For example, Charles Darwin (1809–1882) revolutionized the way humanity thought about itself and challenged long-held beliefs in the measured hand of a divine creator. Psychologist Sigmund Freud (1856–1939) responded to the social and spiritual upheavals unleashed by scientists such as Darwin and political philosophers like Karl Marx (1818–1883). Freud noted that nothing escaped humankind's criticism or resentment, and he believed that the human capacity for destruction outweighed its creativity. After the Enlightenment's optimism, a sense of uncertainty and the need to "play it safe" had set in – despite humanity's ever-increasing ability to understand and shape the world.

POLITICAL REVOLUTION

The 1776 US Declaration of Independence stated that people had the right to abolish any government destructive of life, liberty, and the pursuit of happiness. The French Revolution of 1789 expounded an even more radical vision of human rights, based on a belief that people are inherently free and equal. Everyone should have the right to participate in the legislative process, to make leaders accountable, and to live without fear of arbitrary arrest and punishment. They should also enjoy freedom of speech, opinion, and religion, and equality of taxation. A new political language and framework was born.

CRY FREEDOM
The French Revolution was perhaps the defining moment of political modernity. The genie of equality could never be put back in the bottle.

In February 1792, the revolutionary French government issued a general emancipation of slaves, but ruthless ordinances against vagabondage kept labour on the plantations. In social and economic terms, therefore, little had changed. The French never espoused equal rights for women, but the issue could not be kept off the new political agenda. In 1792, Mary Wollstonecraft's *Rights of Women* articulated the place of the forgotten 50 per cent of society in the struggle for emancipation. In *The Communist Manifesto* of 1858, Karl Marx had already set out the possibility of humanity seizing control of its own destiny and building a new, liberated social order.

EQUAL RIGHTS
Towards the end of the 19th century, suffragettes famously led by Emmeline Pankhurst took up the cause of freedom for women with a passion.

1700

1712 Early slave revolt in North America

1717 Britain, France, and the Netherlands form an alliance to contain expansionist plans of Spain

1720

1724 German philosopher Emmanuel Kant born

1736 Rubber introduced to Europe from Central America

1740

1760

1764 James Hargreaves invents "spinning jenny", England

1770 Luigi Galvani and Alessandro Volta make electricity from chemicals, Italy

1776 US Declaration of Independence

1780

1783 Russia annexes Crimea

1800

1803 US acquires all French territory between Mississippi and the Rockies

ECONOMIC REVOLUTION

In the late 17th and early 18th century, technological changes in farm machinery and experimentation in crop rotation and plant breeding stimulated agricultural productivity. Innovators such as Jethro Tull (1674–1741), famous for his seed drill, were practical people whose passion for innovation owed as much to a desire to commercialize food production as to the new spirit of enquiry. In the late 18th century, the same trend occurred in manufacturing. The Industrial Revolution began in England and spread to Europe, the US, and Japan. New technology and power sources were used in the textile industry. New spinning machines, and the invention of the steam engine, transformed Britain into the workshop of the world. Such technical changes were just part of a much wider transformation of trade, economic relations, and division of labour. In 1851, the Great Exhibition was held in London to show off the newly formed United Kingdom's industrial, military, and economic superiority. Yet the world's leading force soon faced competition. All the great powers were establishing strong financial institutions, and traders around the world were gaining access to the best markets and enjoying the security of operating through contracts.

STEPHENSON'S ROCKET
This groundbreaking design was developed in 1829, 4 years after the first passenger steam train was introduced. Rail travel transformed society and made goods far easier to transport.

IRON BRIDGE
Built in 1779 in Shropshire, the world's first iron bridge dominates an English region that was transformed by iron, steel, and pottery works.

Profile

Simón Bolívar

Born in Venezuela, Simón Bolívar (1783–1830) liberated much of South and Central America from Spanish rule. "El Libertador" conquered the armies of Panama, Peru, Colombia, Ecuador, Bolivia, and Venezuela, and established presidential rule. Bolívar's dream was to unite Spanish America into one republic, but his autocratic rule led to dissent and his "empire" soon crumbled. Bolivia, the country, and the Venezuelan currency (the Bolivar) are still named after him today.

HEIGHT OF PROGRESS
Technological innovation and changes in trade, economic relationships, and division of labour made possible the UK's Great Exhibition of 1851.

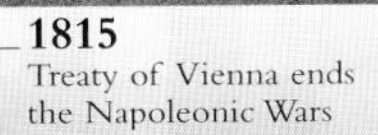

SHIFTING POWER

The reassertion of European monarchies at the 1815 Congress of Vienna quelled aspirations towards national liberty. The war for Greek independence (1820s) was an exception that succeeded in stirring the liberal conscience in Europe – probably because it did not threaten the new European order. A year of revolutions in 1848 again crushed ambitions for national autonomy, but the Italian struggle for independence attracted sympathy and unity was achieved in 1870. Chancellor Bismarck put Prussia at the core of a German empire that covered central Europe by the late 19th century. The major European powers fought for empires to guarantee their raw materials, economic markets, and strategic routes. The "Scramble for Africa" began in earnest in 1885, and within 15 years, the map of Africa was redrawn. The UK no longer dominated global trade and investment: the US had its own growing capitalist class and zone of influence, and Germany started to outpace the UK in iron and steel production. The search for a competitive edge soon took a more military form. Russia, the UK, and Germany each dominated an area of China but were, towards the end of the 19th century, threatened by the presence of the US and a modernized Japan. By 1918, the US brooked no interference from war-devastated Europe, but was not yet powerful enough to stabilize and rebuild the continent. A power vacuum was left at the centre of Europe after Germany had been defeated and Tsarist Russia (guardian of the old order) had become the Soviet Union.

CARVING IT UP
The UK and France divide up the world in this political cartoon. The scale of colonial expansion and the rivalry between the world's leading powers are the crucial features of this era.

WORLD WAR I
The "Great War" was totally catastrophic for Europe, not only in terms of lives lost but also economies destroyed. It left a patchwork of small nation states beset with ethnic rivalries.

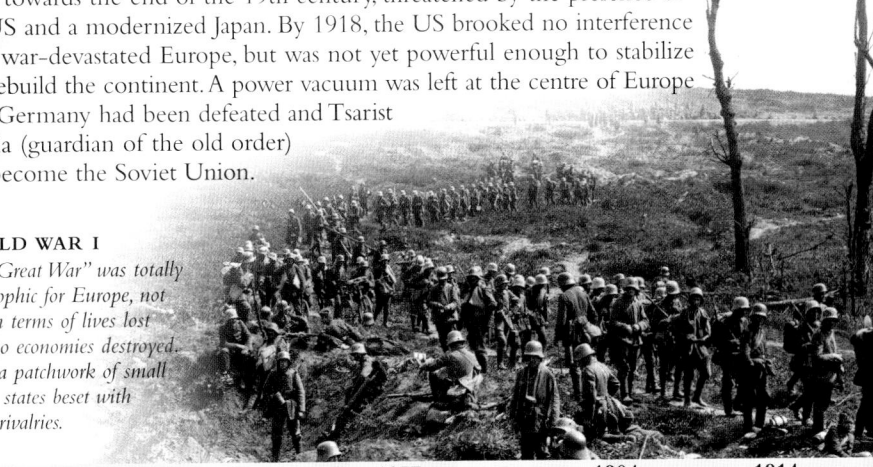

1810 Start of revolutions in Spanish America; by 1826, all Spain's American colonies are independent

1815 Treaty of Vienna ends the Napoleonic Wars

1820

1834 End of Spanish Inquisition (began 1478)

1839 Opium wars begin in China, when Hong Kong is ceded to the UK

1840

1858 Darwin publishes *The Origin of Species*, UK

1860

1861 American Civil War begins (ends 1865)

1877 Famine starts in northern China; kills 10 million in 2 years

1880

1890 First underground railway opens, UK

1896 First modern Olympics, Greece

1900

1904 Russo–Japanese war reinforces Japan's dominance

1914 World War I starts (ends 1918)

1915 Einstein's Theory of Relativity formulated, UK

THE GREAT DEPRESSION

A brief economic boom in the 1920s had little impact worldwide. Then, stock market prices collapsed in the Wall Street Crash of October 1929, and world trade actually shrank. In the resulting chaos, investors withdrew money, companies went bankrupt, and trade crumbled. Economic stagnation and mass unemployment brought hardship and suffering not just to the developed economies, but across the globe. In the US (the world's largest economy), 13 million were unemployed by 1931. Yet the impact was most disruptive in Germany, a society not yet recovered from military defeat in 1918 and revolutionary upheaval in the early 1920s. German economic meltdown provided the perfect context for the National Socialist (Nazi) Party not only to gain power but also to maintain it. Hitler's ruthless desire to carve out a preeminent role for his Third Reich in a new world order was the catalyst for barbarism on a scale previously unknown.

SOUP KITCHENS
Germany suffered a cataclysmic economic collapse following the Wall Street Crash, reducing many people to poverty.

A FATEFUL DAY
Americans took to the streets in panic after the 1929 Wall Street Crash. They withdrew money from investments, companies went bankrupt, and unemployment skyrocketed.

WORLD WAR II

GENOCIDE
In 1941, Jews were told to wear yellow stars. By 1945, 6 million had been killed in Europe.

Within months of becoming German Chancellor in 1933, Hitler opened the first concentration camp. Initially intended for political opponents, these camps became repositories for non-Aryans, especially Jews. The Nuremberg Race Laws formalized longstanding anti-Semitism in 1935, and Jewish persecution culminated in the agreement of the Final Solution in 1942. Genocide was executed with bureaucratic efficiency. The 1939 German invasion of Poland had reopened hostilities between the Great Powers. Unlike in 1914, Germany brushed aside the French army and expelled Britain from mainland Europe. Despite failing in the Battle of Britain (1940), the German High Command risked a Soviet invasion in June 1941. That December, the war became global when Japan bombed Pearl Harbor naval base and the US joined forces with the USSR, the UK, and the colonies. The Soviet Red Army and US economic power crushed Hitler from both sides. War ended in August 1945, with the atomic annihilation of two Japanese cities.

BATTLE OF STALINGRAD
The heroic efforts of the Red Army and the Soviet people helped to ensure German defeat. This battle was the turning point of the war.

ATOMIC BOMBS
The Hiroshima bomb killed tens of thousands instantaneously; radiation accounted for 140,000 lives. The iconic mushroom cloud remains a potent sign of what humanity can do to itself and the planet.

Profile

Nelson Mandela

From 1948, South Africa's Afrikaner government enforced a policy of total discrimination against South African blacks. Nelson Mandela (1918–) was a leading figure in the African National Congress (ANC) during its long-term campaign against the government. Imprisoned for over 27 years, Mandela became a role model for the oppressed. He was eventually released in 1990 and became President of the first ANC government in 1994.

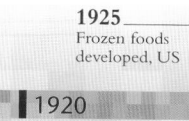

ALTERNATIVES TO CAPITALISM?

A new world order came about with the UN in October 1945. Cold War rivalry between the US, the dominant capitalist power, and the USSR, the dominant communist power, formed the framework of international relations for 40 years to come. Mao's China, outside the main world economy and increasingly distant from the USSR, offered another model for the developing world. The framework of US–Soviet relations regulated international relations. Despite major political crises, such as the Cuban missile crisis in 1962, direct conflict was avoided, and the world economy saw unprecedented expansion and success. Previous capitalist failings, such as economic depression and war, gave credibility to the Soviet model, and many colonial people struggling to build independent states adopted its structures and language. In the 1960s and '70s, it looked like this cultural conflict might lead to a total overhaul of Western society. Although this did not happen, hierarchies were eroded and the role of women changed. Communism fell with the Berlin Wall in 1989, but far from bringing security, the end of the Cold War has made many people feel more insecure.

BREAKING BARRIERS
The tearing down of the Berlin Wall and collapse of the USSR have not brought about the prosperity or political stability anticipated.

1920
- **1925** Frozen foods developed, US
- **1928** First antibiotic, penicillin, discovered by Alexander Fleming, UK
- **1929** Wall Street Crash and start of World Depression

1930
- **1934** Long March begins, China
- **1939** World War II begins (ends 1945)

1940
- **1945** Atomic bombs dropped on Hiroshima and Nagasaki, Japan
- **1947** Mahatma Gandhi achieves Indian independence
- **1949** Apartheid begins in South Africa; Mao declares People's Republic of China

1950

1960
- **1960** First female prime minister elected, Ceylon (now Sri Lanka)
- **1961** Berlin Wall erected

Issue

Global insecurity

Throughout the history of different groups living in close proximity, there has always been a sense of insecurity, however vague. Since the Cold War, many of the more obvious global threats have disappeared, but insecurity remains. Conflict is increasingly defined in terms of terrorism, although this may be no more common today than in the past. Why terrorism has had a huge impact on our collective psyche is debatable – perhaps because it is an ill-defined threat and is carried out by ordinary people.

A GLOBAL ECONOMY

The increasing tendency towards greater economic, social, and technological exchange between countries is known as globalization. Through our technical abilities, we have reduced the size of the world, and this has resulted in wide-ranging economic and cultural changes. The resulting "global village" is unevenly developed and unevenly welcomed. In protest at the increasing emphasis on commerce and profits, many people have adopted new forms of ideology, such as environmentalism, and some question the need to pursue global economic models. Paradoxically, these very environmental concerns have become a new market opportunity for global companies. Globalizers and antiglobalizers do not conform to the old ideological divisions of left and right. Globalizers may simply support capitalist growth or may believe that Western intervention is necessary to bring about global democracy. Equally, they may be rural workers in the developing world who want a road, a dam, or a better strain of wheat. Some antiglobalizers strive to reform or abolish capitalism and US dominance, preserve the planet, or help its more deprived inhabitants. Others are conservatives, wanting to preserve tradition or their own embattled economic and political status. Whether the trend towards globalization is inevitable or desirable will be debated for some time to come.

McDonald's
جنيه
ساندوتش
هامبرجر

ONE WORLD?
Some hope and others fear that the spread of Western culture will result in a homogenous global life experience.

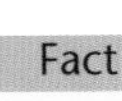

WORLD OF CONTRASTS
Global markets tend to mean abundant choice in Western countries, where supersized meals cause obesity. At the same time, developing countries, such as the Sudan, are unable to feed their populations.

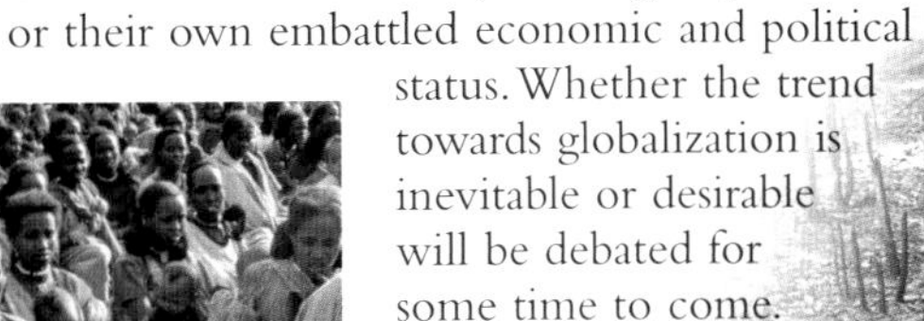

ENVIRONMENTAL DEVASTATION
Many claim deforestation occurs as a direct result of globalization, for example from logging by timber companies. Yet some local farmers also clear land.

Fact

Movement of peoples

Increasing prosperity in the developed world and economic and political problems across the developing world have led many, such as the African immigrants below, to seek new lives in the West. This has led to a new trade in "people smuggling" and growing resentment in developed countries. The perception, however false, that immigrants and asylum seekers gain more benefits than some of the less-advantaged sections of Western society has increased support for rightwing groups through the developed world.

MODIFYING NATURE
We now have the coveted power to alter nature. We can even genetically modify plants and animals – for example, this GM cotton.

TECHNOLOGICAL LEAPS

The 1990s opened with the hope that a new period of stability and progress had begun, but it did not materialize. Debate about humanity's future centres on globalization, US power, the communication opportunities of the internet, and on the scientific possibilities of human intervention in genetic structure. The internet is a recent innovation: in 1989, Tim Berners-Lee produced an internal paper for his European company and a year later received the go-ahead from his boss to write a global hypertext system. Within 6 months he had created the *WorldWideWeb*. The use and size of the Web has grown exponentially. It contains a wealth of accessible data, although people in less-developed regions cannot afford to tap into it. Cheaper travel and mobile phones have also helped to shrink the world. Cultivation of a random mutation of grass led to the advent of bread wheat around 10,000 years ago, a technological leap that seems insignificant now that we are in the process of trying to identify all of our 30,000 genes and determine the sequence of 3 billion chemical base pairs that make up human DNA. We now have a more developed capacity to transform nature than ever before. However, in some ways we may have come full circle in our evolution – whether we allow our newfound power to bring progress or self-destruction is crucial to our future as a species.

SPACE SHUTTLE
Cheap, accessible travel has transformed life for the rich; travel to other planets is even a theoretical possibility.

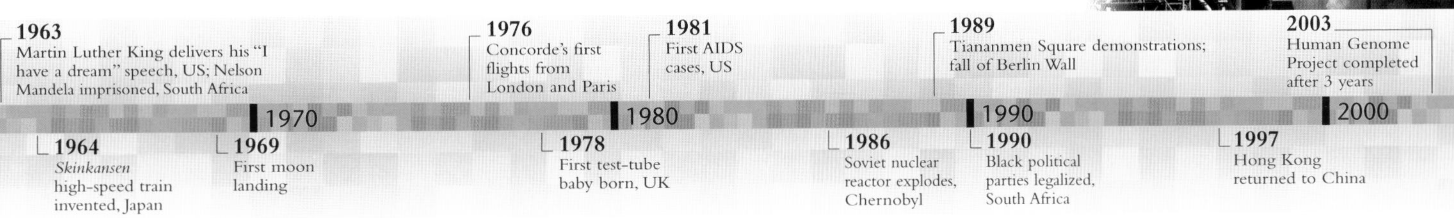
1963
Martin Luther King delivers his "I have a dream" speech, US; Nelson Mandela imprisoned, South Africa
1964
Shinkansen high-speed train invented, Japan
1969
First moon landing
1970
1976
Concorde's first flights from London and Paris
1978
First test-tube baby born, UK
1980
1981
First AIDS cases, US
1986
Soviet nuclear reactor explodes, Chernobyl
1989
Tiananmen Square demonstrations; fall of Berlin Wall
1990
1990
Black political parties legalized, South Africa
1997
Hong Kong returned to China
2000
2003
Human Genome Project completed after 3 years

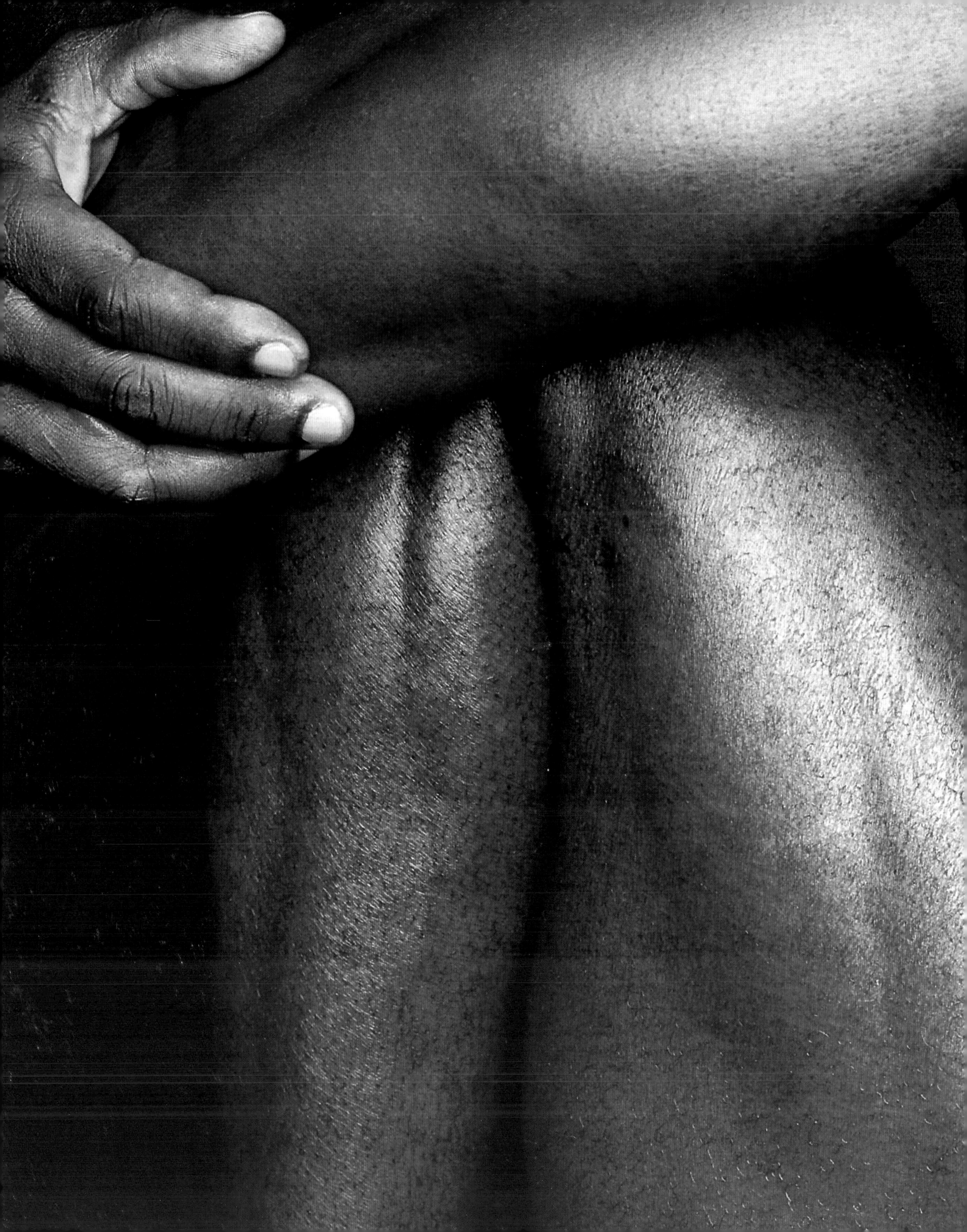

BODY

BODY

Human bodies come in many shapes and sizes. However, with a few differences between the sexes, we are all built according to the same anatomical blueprint. Strip away the skin, of whatever colour, and there will be the same overlapping layers of muscles in the same places. Delve deeper, and the bones are still unmistakably human: the long spine that keeps us standing more upright than other animals; the intricately jointed bones of the hands and feet, and the skull with its flattened frontal planes. One person's internal organs – such as the brain, heart, and intestines – look much like another's, and work to the same internal rhythms.

START OF LIFE
Sperm cluster around an egg cell, trying to penetrate it. If one succeeds, it will fuse with the egg, possibly creating a new human.

We are part of the animal kingdom and belong to a group of animals known as primates, which includes apes (our closest relatives) and monkeys. Primates in turn belong to a larger group of animals – the mammals. In order to understand how humans work, it is useful to see what characteristics we share with other group members and in what important ways we differ.

HUMAN TRAITS

Humans possess the common mammalian characteristics of warm blood and body hair. Together with nearly all mammals, we give birth to live young and produce milk for them. As primates, we are not alone in being able to hold our upper bodies erect (although our firm balance on two legs is unrivalled). The structure and arrangement of the bones in our limbs are similar to those in many of our fellow primates. Like us, other primates have eyes at the front of their faces and stereoscopic vision, and many primates see in colour just as well as we can. However, the human body, which took a few million years to evolve once we branched away from our closest relatives, does have a number of important distinctions. Human hands are specialized for many tasks, such as the skilled use of sophisticated tools; and human feet are modified to bear the entire weight of the moving body. Most significant of all is the human brain, which went on growing during its evolution and is much bigger in proportion to body size than in any comparable species.

HEART FUNCTION
Unlike the ancient physicians, we now know how blood is pumped through the heart.

UNDERSTANDING HUMAN ANATOMY

From the earliest days of civilized society, humans have been curious about what makes their bodies work. For some thousands of years, cutting up bodies to see what goes on inside was severely restricted by the taboos of religion and culture. Theories on human anatomy were, for the most part, based on the dissection of animals and a lot of conjecture. Some early scientists got it surprisingly right, although their ideas were not necessarily accepted at the time. As long ago as 500BC, the Greek physician Alcmaeon suggested that the brain was the seat of our thoughts and emotions. On the other hand, mistaken ideas were rife, most famously those of Galen, a Greek-born doctor who practised in Rome during the 2nd century AD. Galen's beliefs that blood is made in the liver and that a flame burning in the heart keeps us alive were just two of the theories that stayed unchallenged for centuries. Not until scientific knowledge took a leap forward in the Renaissance were some of the truths about the human body revealed. The observations of the Flemish doctor Andreas Vesalius (1514–1564), who published the first accurate illustrations of human anatomy, were the start of a new age of understanding. With each century, fresh discoveries were made. In 1628, William Harvey proved that blood circulates around the body; in 1775, the French chemist Antoine Lavoisier found out that cells use oxygen for fuel; British surgeon William Bowman first described the function and structure of the kidney in 1842; hormones were discovered in the early 1900s. So, gradually, the pieces of the human jigsaw were slotted into place. The growth of knowledge about the body accelerated as the 20th century passed, and is continuing unabated in the 21st century.

THE MODERN BODY

In countries where there is an adequate food supply and good standards of basic hygiene, the average human body is taller and longer-lasting than it was in previous centuries. With a

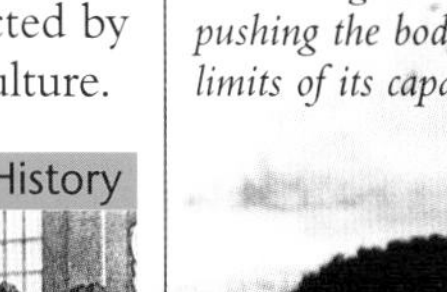

TESTING THE BODY
Like this young break-dancer, humans are constantly testing their strength and agility, pushing the body to the limits of its capabilities.

History

Anatomy lesson

Dissections of dead bodies were once carried out more as elaborate stage performances than as anatomy lessons. Many of the audience were merely curious onlookers rather than medical students. Until the 19th century, the bodies used were nearly always those of executed criminals. In this 1751 engraving, the English artist William Hogarth caricatures a public demonstration of the anatomist's craft.

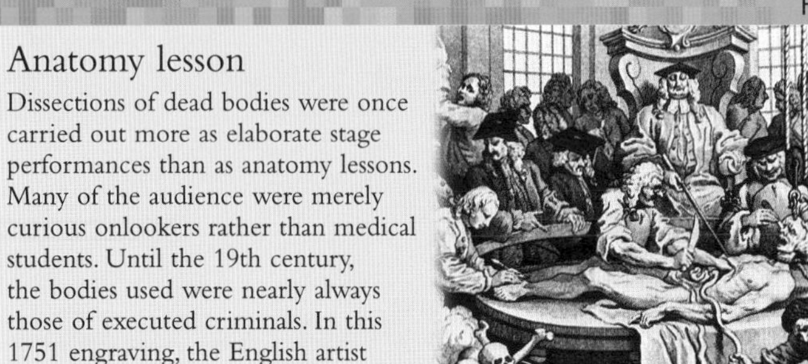

DEFENCE CELLS
These cells are lymphocytes, which belong to the body's disease-fighting system and are found mainly in the blood. They have just formed by division of a single cell.

better understanding of health and body functioning, people are learning how to keep themselves in good order over a lifetime. Although humans in modern society tend to lead sedentary lives, we have thought up ways of meeting the body's need for regular physical activity. Many communities offer a wide choice of sports and exercise facilities; and some people enjoy the challenge of exploring the amazing capacity the human body has for physical improvement. The limits to which the body can be pushed are being extended all the time. Every year, new records of speed, strength, and stamina are set by athletes, gymnasts, and sportspeople. These achievements demonstrate the body's potential for increased lung power, more efficient use of the heart, stronger muscles, and greater flexibility. The modern body is under our control in other ways. If we are not satisfied with our appearance or shape, there are huge industries devoted to making the body thinner, more attractive, or younger-looking. We can even limit our fertility, making informed choices about whether or not to produce children. At the same time, we can also improve our chances of creating new life.

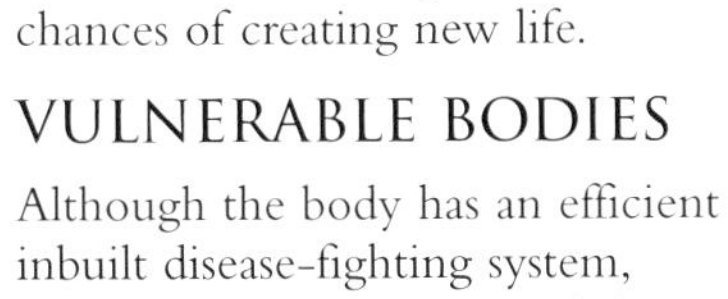

VULNERABLE BODIES

Although the body has an efficient inbuilt disease-fighting system, from time to time it needs help. The enormous range of drugs available in the developed world provides invaluable back-up protection and can often cure diseases that get past our natural defences.

Modern medicine has wiped out or greatly reduced diseases such as smallpox or polio that once killed people in their millions worldwide. It is possible, too, to replace some body parts, such as the heart, liver, and joints, when the originals are diseased, damaged or worn out. Despite these great advances, the human body is still vulnerable. Old threats may have been overcome, but we constantly face new ones, some of our own making. In all affluent societies, illnesses caused by overeating and inactivity are increasing. We create chemicals to destroy diseases in the food-chain and to boost food production, but some of these chemicals could be gradually poisoning our bodies. In some areas, environmental pollution is thought to be a contributing factor in the increase of cancers and birth defects. Infectious organisms are becoming resistant to drugs, and are more dangerous than before. New diseases, such as HIV/AIDS and SARS (severe acute respiratory syndrome), are emerging to which we have no natural resistance. The scale and speed of international travel means that once-localized diseases are carried to new areas within a few hours.

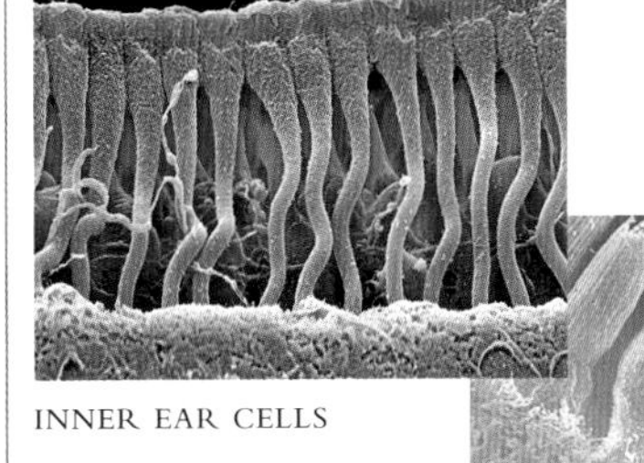

INNER EAR CELLS

RETINAL CELLS

UP CLOSE
The elaborate structures of cells in the inner ear (above) and in the retina of the eye (right) are revealed by an electron microscope.

NEW DISCOVERIES

Doctors can now look at the innermost recesses of the living body without cutting it open. There are techniques for viewing the body that were unimaginable a few decades ago. For many years X-rays, discovered in 1895, were the only way to see inside the body without surgery – and only hard tissue, mainly bone, was shown. Newer imaging techniques enable soft tissues such as muscle to be seen, and can highlight hollow organs. It is also possible to look at tissue activity, to find out how well an organ is functioning (*see* Looking inside the body, below). Another modern technique is ultrasound, which uses sound waves and is a safe method for monitoring babies in the womb. Major advances have produced 4-D ultrasound, which constructs moving images. The examination of hollow organs is often carried out by endoscopy, in which a tube-like optical instrument is inserted through a body opening, such as the mouth or anus. One of the most exciting inventions is the electron microscope, which is used for examining samples of body tissue in minute detail. With a magnifying power thousands of times greater than an optical microscope, it facilitates the study of the internal structure of any cell in the body.

Fact

Looking inside the body

Modern imaging techniques provide reliable pictorial information about the inner body. In computerized tomography (CT) scanning a series of X-rays passes through the body, producing cross-sectional images ("slices"), which are displayed on a monitor. MRI scanning uses radio waves combined with magnets to pick up tiny signals from atoms in the body. These signals are built into images by a computer. Positron emission tomography (PET) scanning is a technique that detects the body's uptake of injected radioactive substances. PET produces colour-coded images showing cell activity.

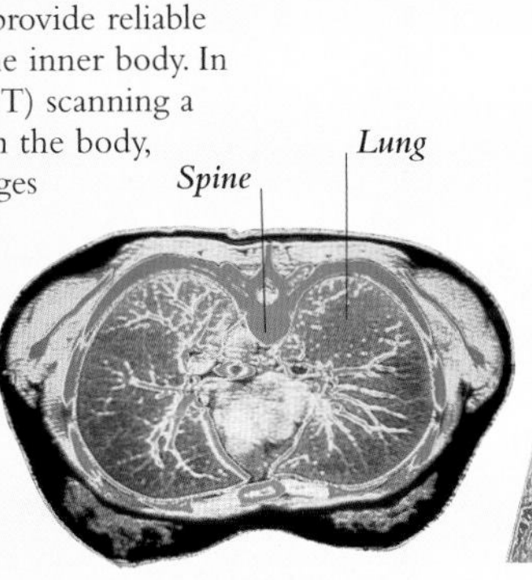

CT SCAN
This "slice" is a cross-section of the chest, which shows tissues of different densities, including the lungs (red) and bones (green).

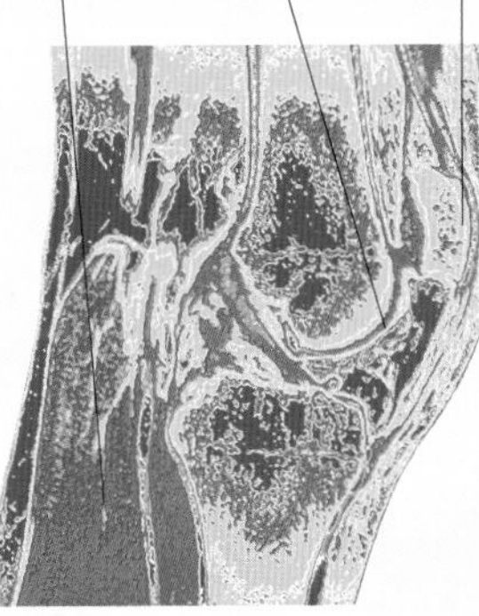

MRI SCAN
An MRI scan of the knee gives a detailed picture of the different structures in the joint, including the bones, muscles, and cartilage.

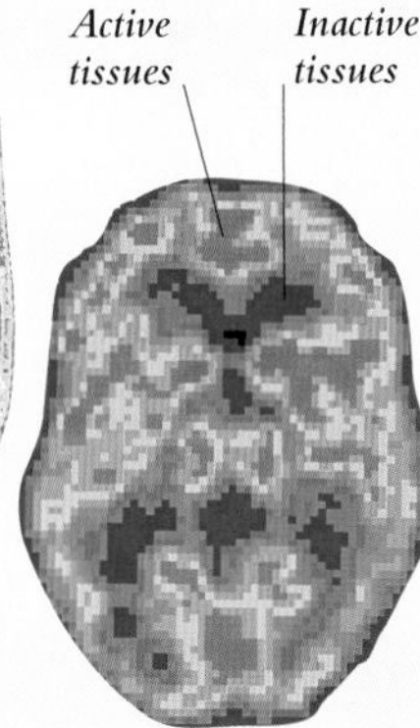

PET SCAN
Radiation detected in this PET scan of the brain shows areas of active (red) and inactive (blue) tissues.

BODY

CONNECTIONS

The body is a series of connecting parts. This tangle of abdominal arteries links many areas (including the kidneys, spleen, liver, and legs) to the body's blood supply.

BUILDING A BODY

Construction of the human body starts at microscopic level with cells, the basic units of all living things. Cells of the same type combine to make tissues, the body's materials; and collections of tissues make organs, the working parts of the human machine. A series of organs performing one of the body's major processes or functions, such as digestion or circulation of the blood, makes a system. To function properly, each system is reliant on the others and all of them interconnect to make the complete body.

The body's basic ingredients are thousands of different types of chemical substances that are similar to the substances found in food: for example, fats, proteins, and minerals. Each chemical consists of units known as molecules, and each molecule is constructed from combinations of tiny particles called atoms. All our body cells contain specialized working parts, known as organelles, that are made up from chemical molecules.

THE DOUBLE HELIX

The whole of the body, from the organelles inside cells to the tissues, organs, and systems that keep us alive and functioning, is created under the instructions of a single chemical: DNA (deoxyribonucleic acid). Molecules of DNA are coiled tightly into tiny threadlike structures called chromosomes, which are packed into the nucleus (central unit) of each cell. The long DNA molecule is shaped like a twisted ladder with the rungs made of building blocks called nucleotide bases, which are arranged in specific pairs. Sections of DNA make up genes and these have two functions. Genes provide cells with the instructions they require for making proteins and other molecules that control cellular processes – elements that are needed for the development and growth of all the body's organs and structures. Genes are also the means by which physical and some mental characteristics are passed on from generation to generation. For instance, there are combinations of genes that are responsible for creating the colour of our eyes and hair or the shape of our noses. When a body cell divides in order to make a copy of itself – in the process of growth and repair that occurs naturally all over the body – the DNA in the cell is duplicated as well. This ensures that every new cell that is created contains a complete set of instructions for the human body.

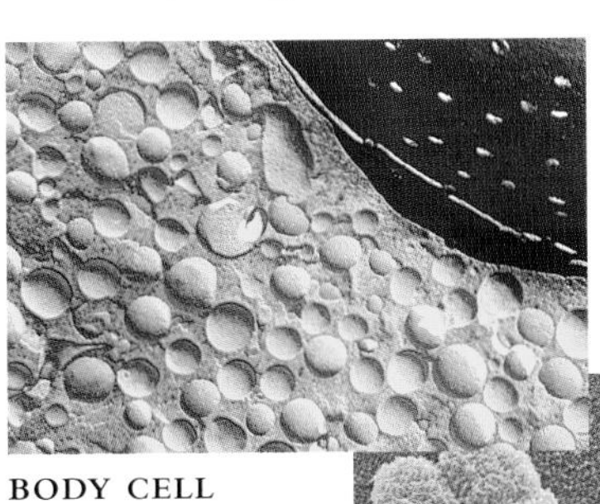

BODY CELL
The nucleus of a body cell (dark area) contains genetic instructions. The cytoplasm (green) that surrounds the nucleus contains structures needed for cell function.

DNA
These structures are chromosomes, coils of DNA (genetic material) found in cells. Normally in single strands, chromosomes join in an X-shape when cells divide.

LIVING CELLS

We have trillions of cells of various types in our bodies, and each one is a world in its own right. Cells are too minute to be seen with the naked eye but they can be studied under a microscope. The degree of magnification possible with modern instruments reveals the astonishing inner world of cells. These tiny living organisms are crammed with working parts that keep the body functioning. The cell nucleus houses our genetic material, DNA, while precisely arranged in the fluid that fills the rest of the cell are other essential components such as mitochondria, where biochemical energy is created. Although cells have many features in common – for example, all cells break down glucose to use for energy – they also have specialized roles and are often highly distinctive in appearance. For example, there are certain cells, neurons, that carry electrical signals to and from the brain. Neurons are easily identified by their long "wires", axons, down which the signals travel. The function of red blood cells is to transport oxygen around the body. The red coloration and dimpled shape of these blood cells make them unmistakable. The lifespans of cells vary. Some kinds of cell survive for a few hours, others last a lifetime.

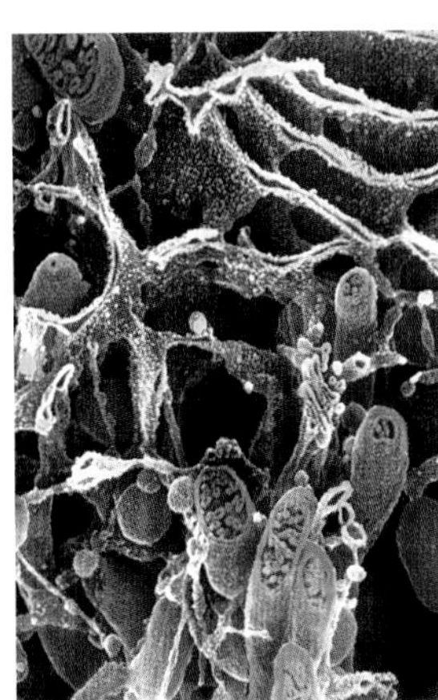

CELL ENGINES
This photograph shows the interior of a body cell. The red structures are mitochondria, the driving forces of the cell, which turn food into energy.

BODY BUILDING

Isolated cells are fragile and can build a body only when they join together to form tissues. There are various types of tissue in the body, – including muscle, the nervous tissue found only in the brain and spinal cord, and the connective tissues that bind everything together. These tissues all have different roles in the maintenance of our body structures and the functioning of our organs. The constant replacement of individual cells in any one mass of tissue keeps everything in good repair. For tissues to function efficiently, they need a support system that provides a supply of nutrients, and a communication system that keeps the tissue in touch with the rest of the body. Tissues have a direct blood supply that provides nutrients through a network of tiny vessels, and also removes waste products generated by cell action. If the blood supply is interrupted, the tissue may die. Most tissue is also threaded by nerve fibres so that messages pass to and from the brain (which is how we feel pain).

Tissues do not perform their functions in isolation. It takes various types of tissue to make a working body organ, such as the heart, stomach or kidneys. For example, muscle tissue may provide the organ's movement (as in the intestines), while other types of tissue produce mucus to provide lubrication and protection.

Like cells and tissues, organs cannot operate alone but must instead be integrated with other organs to build body systems. Examples of organ groups are the respiratory system, which brings oxygen into the body, and the urinary system, which eliminates waste products from the body.

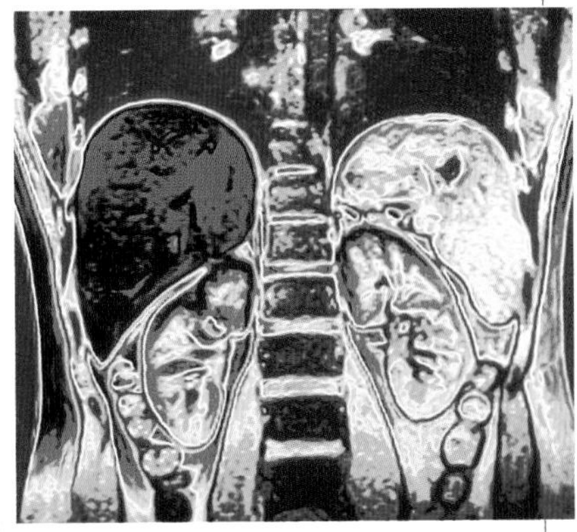

VITAL ORGANS
The liver (dark structure on left) and the kidneys (seen on either side of the spine) are among the most vital parts of the body's working machinery.

Fact

Tissue engineering

It is possible to create tissues in the laboratory for replacing damaged parts of the body. Some replacement tissues are entirely synthetic. Others are created by "seeding" a chemical compound with human cells that are then stimulated into growth.

NEW TISSUE
This bubble is replacement tissue for the cornea (front of the eyeball); it was cultured in the laboratory from cells.

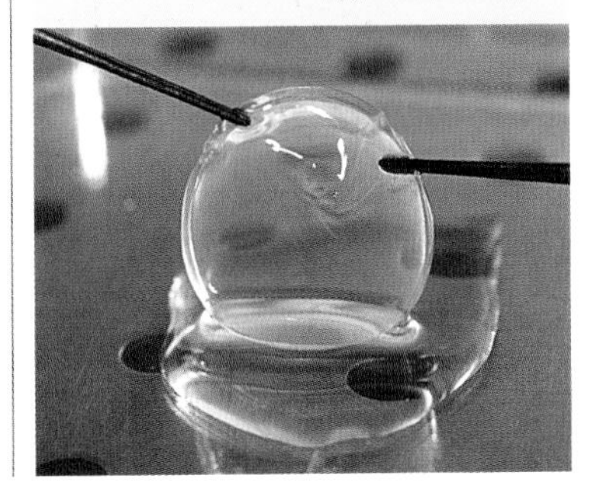

DNA

The chemical DNA (deoxyribonucleic acid) is found within the nucleus of every cell, where it is spiralled up tightly into structures called chromosomes. DNA is the master chemical in the body – the "key" to all life. It contains the recipe for making proteins, which are needed for the development and growth of organs and structures. DNA is the basis for inheritance: the information that enables characteristics to be passed down the generations is carried by genes, which are made of DNA.

MAKING PROTEINS

Some proteins in the body make up structures such as skin; others are hormones or enzymes that control cell activities. A major function of DNA is to provide the instructions for making proteins. DNA is made of molecules called bases (*see right*) whose arrangement provides the template for assembling proteins. Proteins are made from amino acids. Instructions held by DNA for assembling amino acids are relayed by a chemical called messenger ribonucleic acid (mRNA) in stages called transcription and translation.

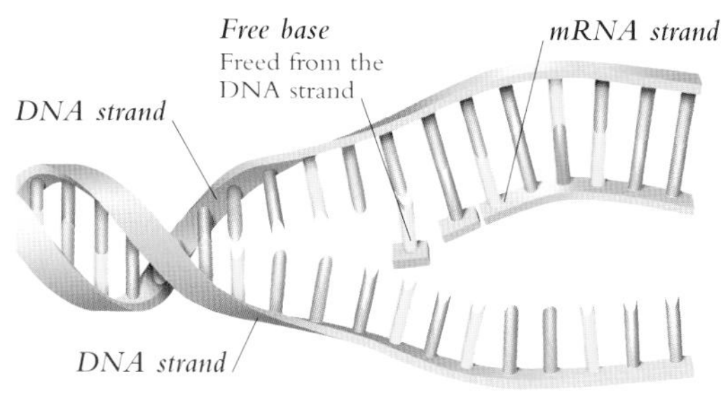

TRANSCRIPTION
Strands of DNA separate along a portion of their length. Free bases attach to bases on one DNA strand to make mRNA. The mRNA carries the instruction for making a protein and moves into the cytoplasm surrounding the nucleus of the cell.

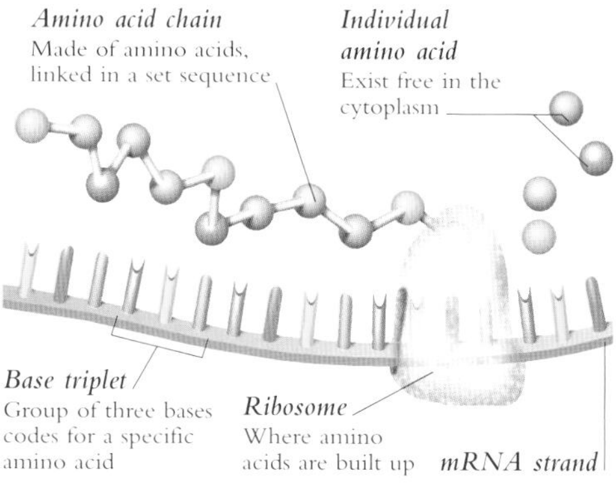

TRANSLATION
A structure called a ribosome moves along the strand of mRNA three bases at a time. The ribosome attaches specific amino acids in place according to the sequence of bases in the mRNA triplets.

COMPLETED PROTEIN
When the ribosome reaches the end of the mRNA strand, it detaches itself from the assembled chain of amino acids. The chain then folds up to form the newly completed protein.

Mitochondrion Structure containing a small amount of DNA

Cell

Nucleus Contains most of the body's DNA in the form of chromosomes

Centromere The point at which the chromosome splits when a cell divides

Chromosome Normally rod-shaped, chromosomes become X-shaped when a cell divides

Supercoiled DNA Chromosomes coil up very tightly, becoming thicker and shorter, before a cell divides

STRUCTURE OF DNA
DNA resembles a twisted ladder, or double helix, its sides made of linked phosphate and deoxyribose (sugar) molecules. The rungs are molecules called nucleotide bases – adenine, guanine, cytosine, and thymine – that are linked together in specific pairs.

Histone DNA spools around eight of these proteins, which regulate the process by which DNA makes proteins

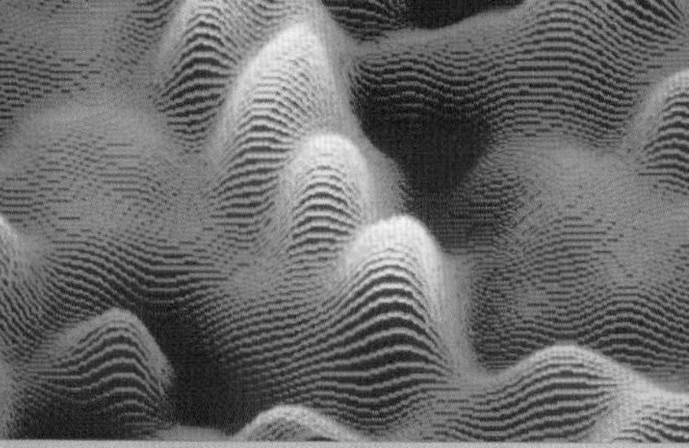

SCAN OF DNA
The yellow "peaks" on this specialized microscope picture of DNA correspond to the coiled ridges of the DNA's two intertwined strands: its double helix.

DNA backbone The two rungs that form the backbone of DNA are made from sugar and phosphate molecules

History

Watson and Crick

In 1953, scientists James Watson and Francis Crick – who were working at a laboratory in Cambridge, Britain – discovered the structure of DNA, using Chargaff's ratios of the bases of DNA and the X-ray crystallography of fellow scientists Maurice Wilkins and Rosalind Franklin. They built a model of the DNA molecule, showing the form of a double helix, with which they explained their findings. Watson, Crick, and Wilkins were awarded the Nobel prize in 1962, Rosalind Franklin having died of cancer in 1958.

REPLICATION OF DNA

Body cells are dividing continuously, during periods of growth and to compensate for cell damage. Before a cell can divide to make new body cells (a process called mitosis, *see right*) or egg and sperm cells (a process called meiosis, *see* p138) the DNA contained in the cell must be replicated, or copied. This process is possible because strands of DNA are able to "unzip" themselves along their length and separate. Each of the two strands in the original DNA acts as a template against which two new strands are built.

STAGE ONE
The original DNA double helix splits open at several points along its length. This process produces areas where there are two separate single strands.

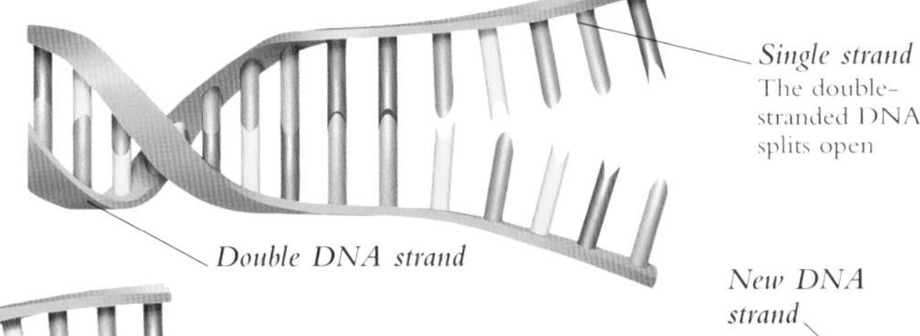

STAGE TWO
Free bases (from the DNA strand) are attached to both of the single strands of DNA. The order in which free bases join to the single DNA strands is determined by the DNA bases that are already present on the single strand.

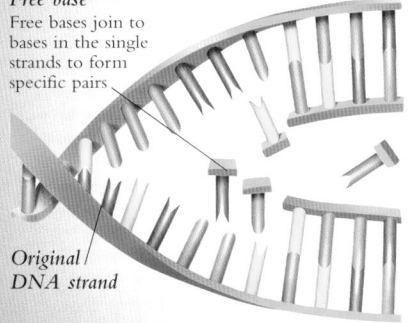

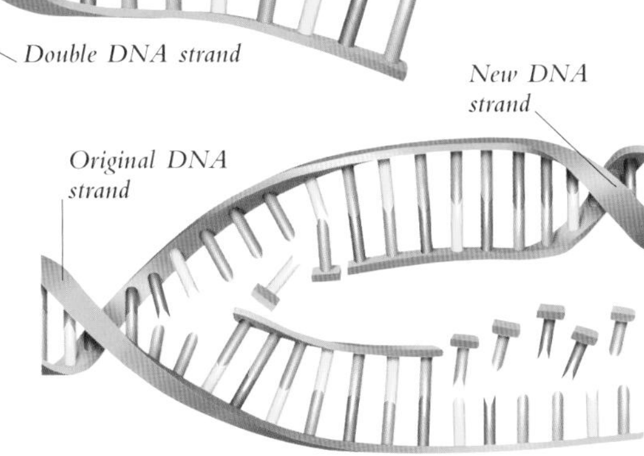

STAGE THREE
While the bases attach to the strand, each of the two newly formed double strands starts to twist. The process continues along the whole length of the DNA, eventually producing two identical double DNA strands.

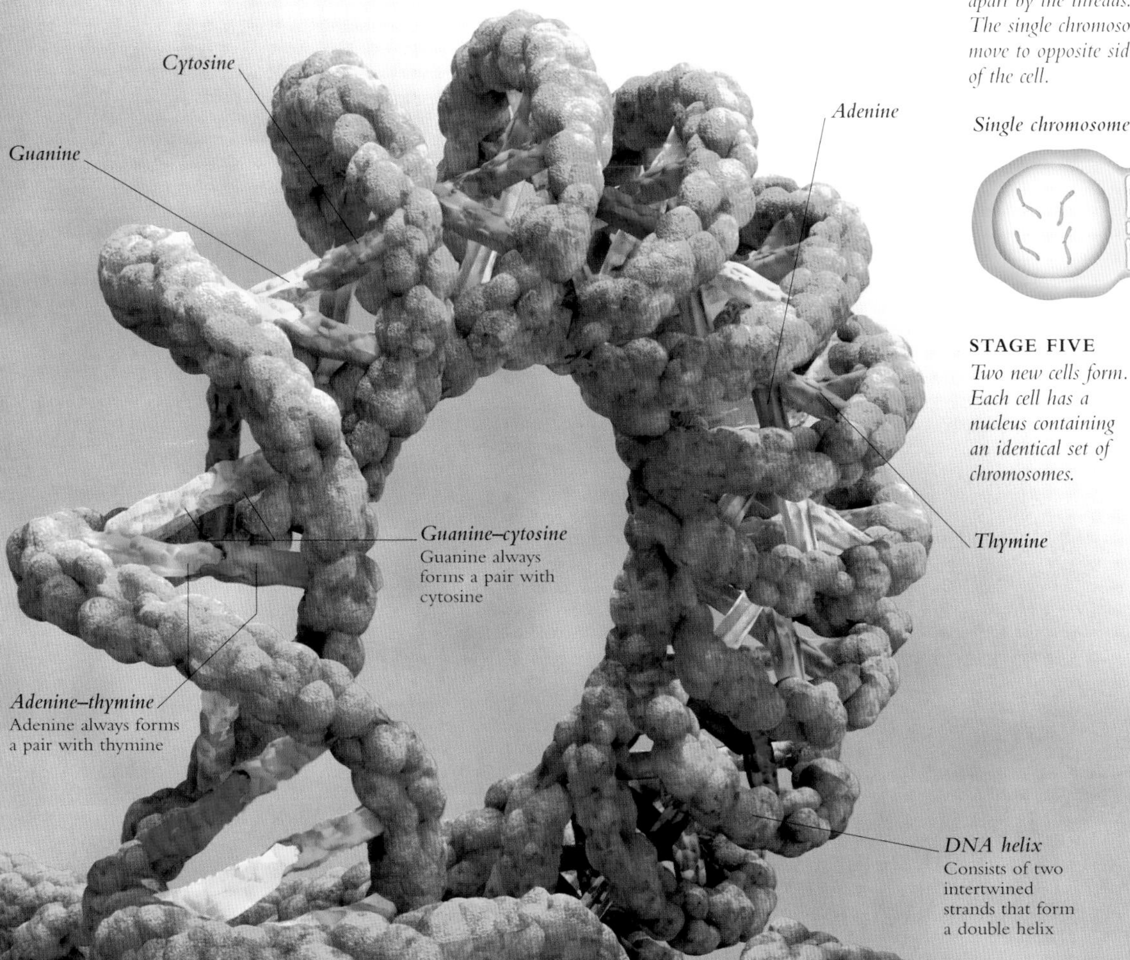

Guanine always forms a pair with cytosine

Adenine–thymine
Adenine always forms a pair with thymine
DNA helix

MAKING NEW BODY CELLS

Before new body cells can be made, in a process called mitosis, DNA must first be replicated, or copied (*see left*). The double chromosomes of duplicated DNA can then line up and split, forming identical single chromosomes. The resulting cells are "daughter" cells, identical to the original cell, and these new cells contain the full complement of chromosomes. This process happens to the full set of 46 chromosomes that exist in each cell, but, for simplicity, only four chromosomes are shown in this representation of the process (*see below*).

STAGE ONE
The DNA in the chromosomes is copied to form two identical strands joined in the centre by a structure called a centromere.

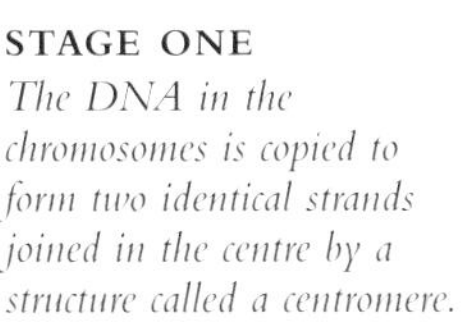

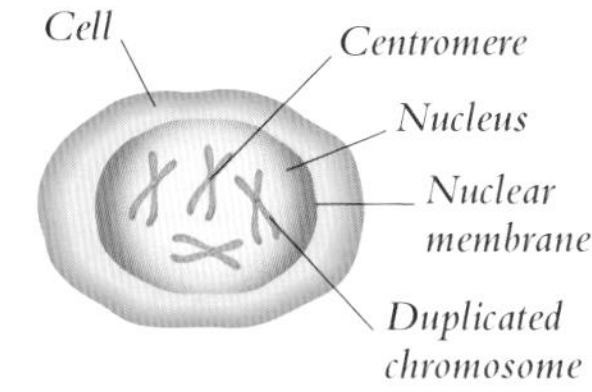

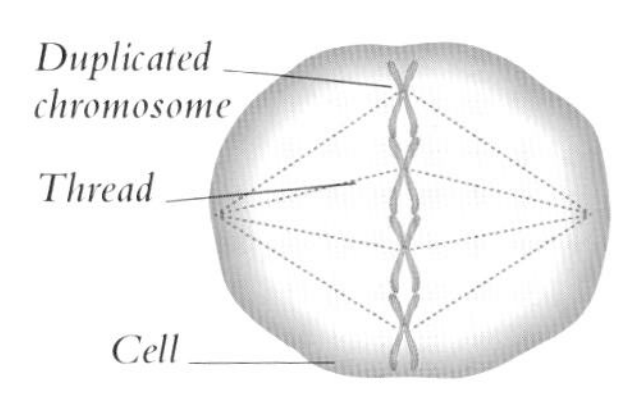

STAGE TWO
The membrane around the nucleus breaks down and threads form across the cell. The chromosomes line up on the threads.

STAGE THREE
The duplicated chromosomes are pulled apart by the threads. The single chromosomes move to opposite sides of the cell.

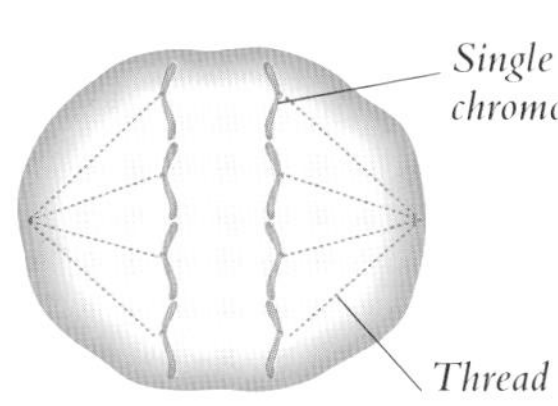

STAGE FOUR
A nuclear membrane forms around each set of single chromosomes. The cell begins to divide into two cells.

STAGE FIVE
Two new cells form. Each cell has a nucleus containing an identical set of chromosomes.

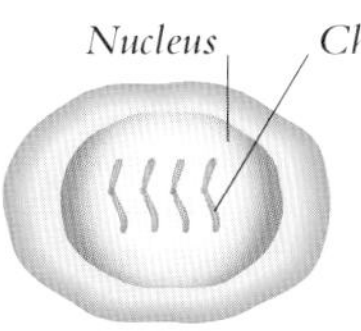

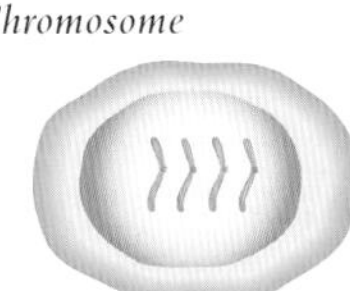

Fact

Mitochondrial DNA

Mitochondria (energy-producing structures in cells) contain a small amount of DNA (shown in red, below). Unlike DNA in the cell nucleus, which is inherited from both parents, mitochondrial DNA is inherited only from the mother. Mistakes in copying mitochondrial DNA is the cause of a small number of genetic disorders.

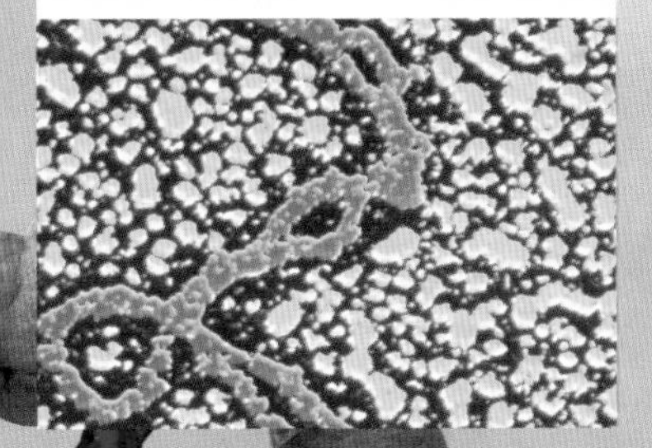

CELLS TO SYSTEMS

The living structure of the human body is built from different types of cells. Cells mass together in clumps or layers to form tissues, which are specialized for different roles within the body. Some tissues form linings – for organs, for example – while others provide a framework, bind body parts together, act as insulation, or allow movement. Two or more tissues make up organs like the heart or the stomach, the working parts of the human machine. When a series of organs functions as a team to facilitate one of the body's major processes, they comprise a system, such as the digestive or nervous system. These systems themselves cannot function alone, but link up to and interact with other systems to form a living, working body.

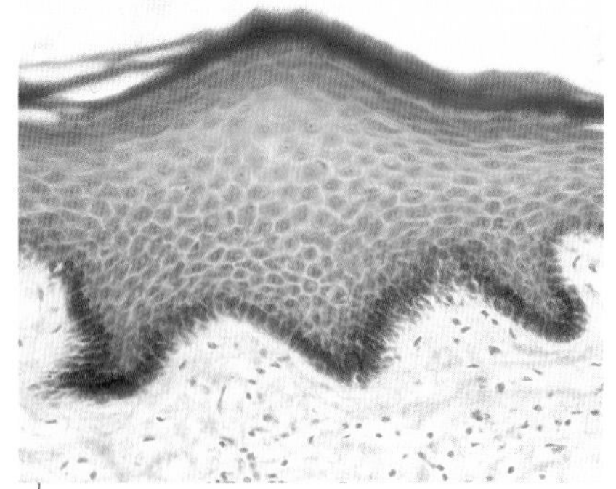

EPITHELIAL TISSUE
The top layer of skin (coloured area above) is composed of a type of tissue (made of epithelial cells) whose main function is to act as a protective outer layer for the body.

UNITS OF LIFE

The cell is the fundamental unit of life. We all come into existence as a single fertilized cell that rapidly multiplies. By the time we reach adulthood, however, the cells that make up the body number approximately 75 trillion. Each of these tiny structures, visible separately only under a microscope, is packed with working components. Cells also contain DNA (*see* p52), the genetic material that is responsible for our development and our individual characteristics. Some cells last a lifetime; others wear out after a day or so, or fall victim to damage or disease. Every day, the body manufactures replacements (*see* Repair and replacement, p104). Although each cell is self-contained, it does not work in isolation. Cells act together in an organized way in order to function effectively – they communicate with one another via chemical messages.

Cytoplasm
A jelly-like fluid in which the organelles float; it is mostly water, but also contains enzymes, amino acids, and other molecules needed for cell function

Cytoskeleton
The internal framework of the cell, made up of thread-like filaments

Peroxisome
Sac where some enzymes are produced and some cell substances oxidized

Lysosome
Contains powerful enzymes that break down harmful materials and dispose of any unwanted substances

Ribosomes
Small granular structures, here attached to the endoplasmic reticulum, that play a key role in the assembly of proteins

THE VARIETY OF CELLS

Around 200 different types of cell have been identified in the human body. Cells begin to differentiate, becoming specialized for particular roles, at a very early stage in our development. Under the influence of growth chemicals, the primitive stem cells of an embryo pass rapidly through various stages before reaching their destinies, for example, as red blood cells, nerve cells, or muscle cells (*see below*). Cells may differ widely in their external appearance and in their activities, but almost all of them share the same internal components (although red blood cells are exceptional as they do not have a nucleus). They also have many basic functions in common. For example, most cells play a part in generating energy from food, a process that keeps our organs in full working order and, ultimately, keeps us alive.

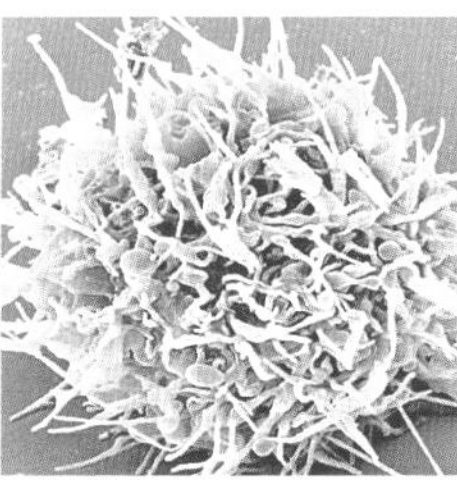

STEM CELL
It is from the stem cells in the bone marrow that all blood cells originate.

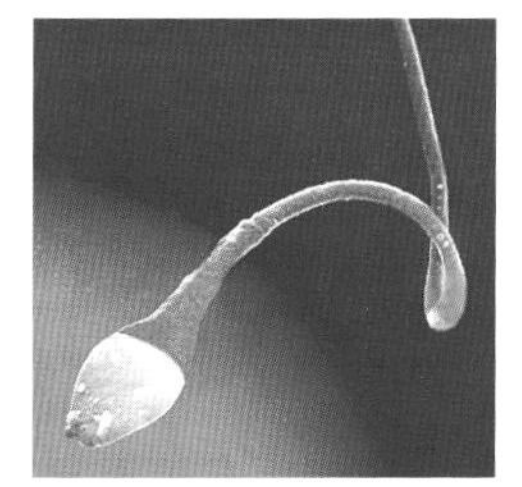

SPERM
The whip-like tails of sperm cells propel them up the female genital tract.

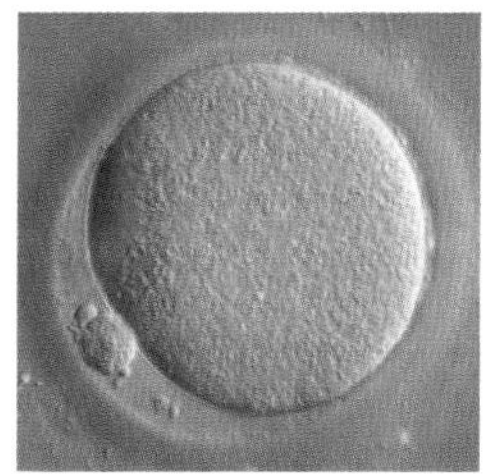

OVUM (EGG CELL)
Egg cells have a protective membrane, which grows thicker after fertilization.

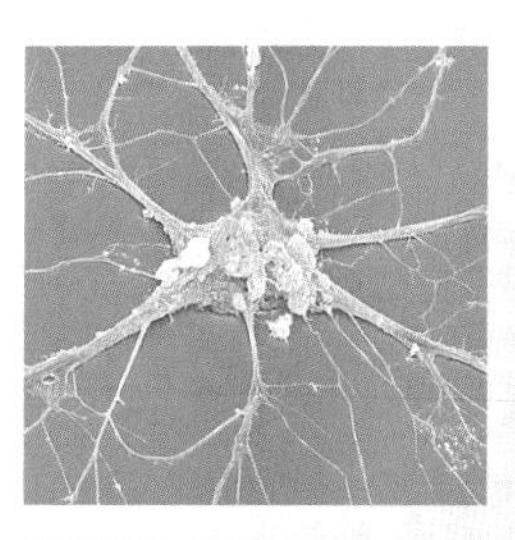

NERVE CELL
Signals between nerve cells pass between long filaments called axons.

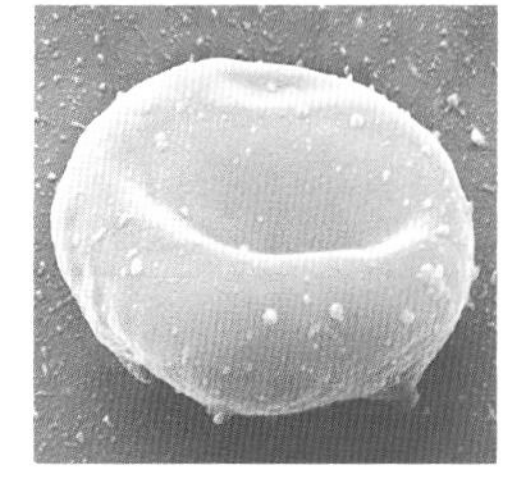

RED BLOOD CELL
These pigmented cells give blood its colour. They carry oxygen around the body.

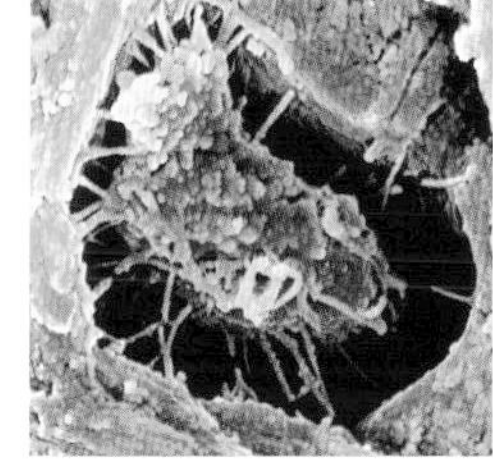

OSTEOBLAST
This type of bone cell lays down calcium to maintain bone strength.

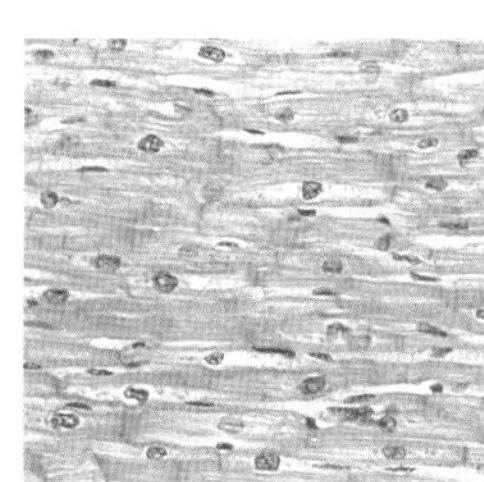

MUSCLE CELL
These cells can alter their length, which varies the force of contraction.

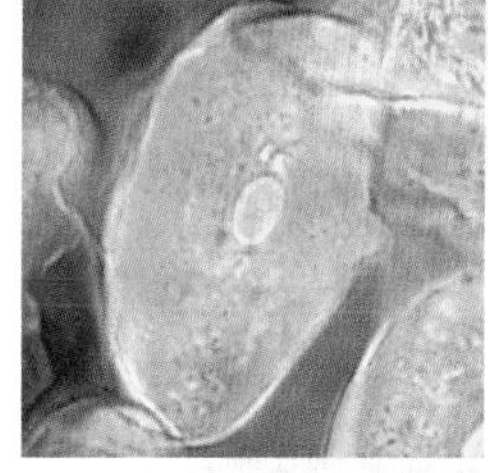
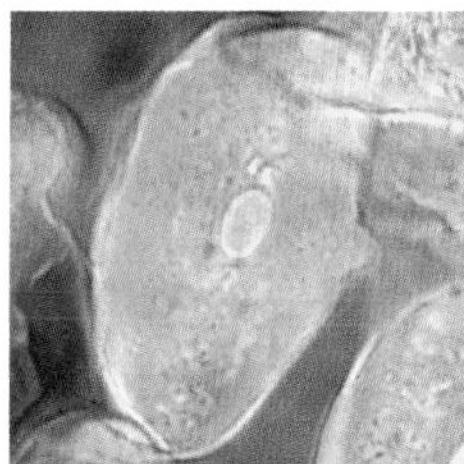

EPITHELIAL CELL
Various types of epithelial cell form skin and tissues that line or cover organs.

Fact

Red blood cells

An average man has approximately 25 trillion red blood cells in his body – a third of the estimated total number of cells in the human body. These curved, disc-like cells have a large surface area, maximizing oxygen absorption from the lungs, but are flexible enough to squeeze into small blood vessels to transport the oxygen all over the body.

BODY

Centriole
Structure located near the centre of the cell that play a key role in cell division

Vacuole
Sac that transports and stores ingested materials, waste products, and water

Mitochondrion
Contains a small amount of DNA and produces adenosine triphosphate (ATP), which provides energy needed for many cell functions

Nucleus
Contains most of the cell's DNA (genetic material), in the form of chromosomes

Pore
Allows substances to pass in and out

Smooth endoplasmic reticulum
Helps to transport materials through the cell; also breaks down toxins and is the main site of fat metabolism

Rough endoplasmic reticulum
Helps to transport materials through the cell; also site of attachment for ribosomes, which are important for the manufacture of proteins

Golgi complex
A stack of flattened sacs that receive protein from the rough endoplasmic reticulum and repackage it for release at the cell membrane

Vesicle
Sac containing substances such as hormones or enzymes, which are secreted at the cell membrane

Nucleolus
Region at the heart of the nucleus; plays an important role in ribosome production

Cell membrane
Outer membrane enclosing the cell and regulating the entrance and exit of substances

History

Identifying the cell

English physicist Robert Hooke coined the term "cell" in his ground-breaking book *Micrographia*, published in 1665. The book outlined his observations of a variety of organisms under a microscope that he devised. The box-like cells visible in this slice of cork reminded him of the cells of a monastery.

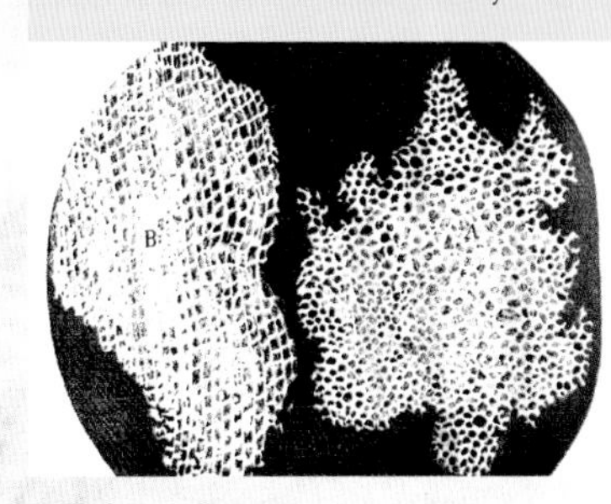

CELL FUNCTION

Cells are busy factories, carrying out several thousand different tasks in an orderly, integrated way. Each type of cell has a specialized role, but most cells are involved in the breakdown of glucose, creating energy to drive their respective activities. Small structures called organelles within each cell carry out the cell's vital activities, coordinated by the cell's control centre, the nucleus. One important task of organelles is the production of proteins, which are needed to carry out vital biochemical reactions in the body as well as for development and growth. Some organelles are involved in digestion processes while others can destroy dangerous chemicals that may potentially harm the cell.

FEATURES OF CELLS
Most of the cells in the human body contain smaller substructures called organelles ("little organs"), each of which performs a highly specialized task. Enclosing the cell contents is a membrane that regulates the passage of substances into and out of the cell.

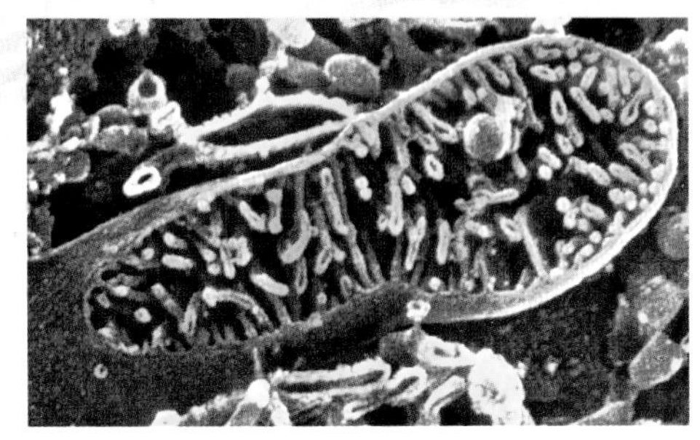

MITOCHONDRION
The cell's powerhouse, this organelle is the site of respiration and the breakdown of fats and sugars for energy. Inner folds contain energy-producing enzymes.

TYPES OF TISSUE

Collections of similar cells, and the substances around them, form tissue. There are five main types of tissue: epithelial, connective, skeletal, nervous, and blood. Epithelial tissue lines and protects body organs. It is fairly closely packed and occurs in sheets that may be several layers thick. Connective tissue is the supporting tissue of the body, and includes bone, cartilage, and fat. It is more loosely packed than epithelial tissue. The protein collagen is very important in giving connective tissue its toughness. There are three types of muscle tissue. Skeletal muscle is attached to bone and allows us to make voluntary movements; smooth muscle tissue is involved in involuntary movements; and cardiac muscle is a specialized tissue allowing the coordinated beating of the heart. Nervous tissue consists of nerve cells (also known as neurons) that conduct electrical impulses, and supporting cells, called glial cells, which supply the working nerve cells with nutrients and oxygen.

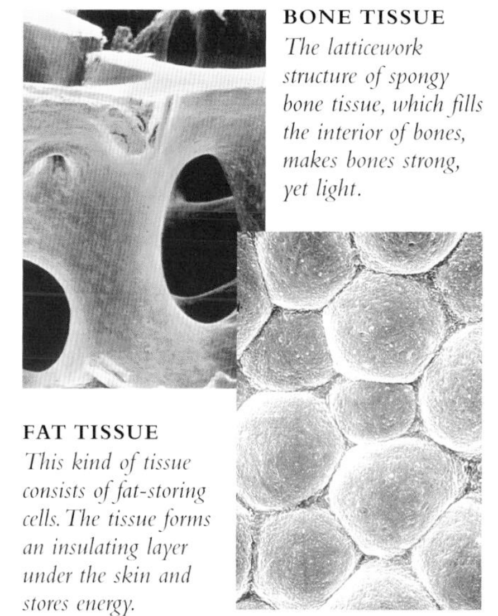

BONE TISSUE
The latticework structure of spongy bone tissue, which fills the interior of bones, makes bones strong, yet light.

FAT TISSUE
This kind of tissue consists of fat-storing cells. The tissue forms an insulating layer under the skin and stores energy.

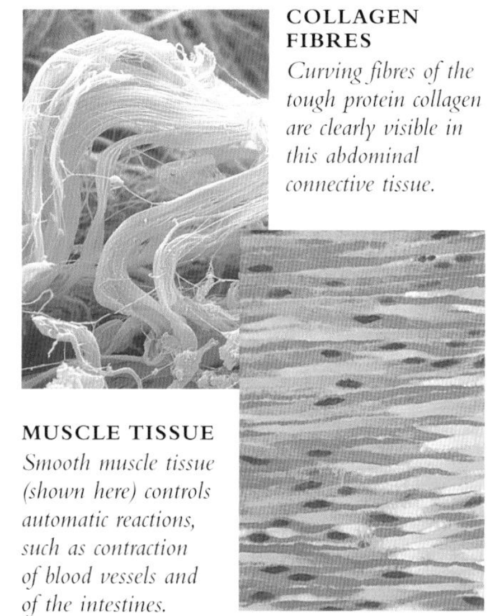

COLLAGEN FIBRES
Curving fibres of the tough protein collagen are clearly visible in this abdominal connective tissue.

MUSCLE TISSUE
Smooth muscle tissue (shown here) controls automatic reactions, such as contraction of blood vessels and of the intestines.

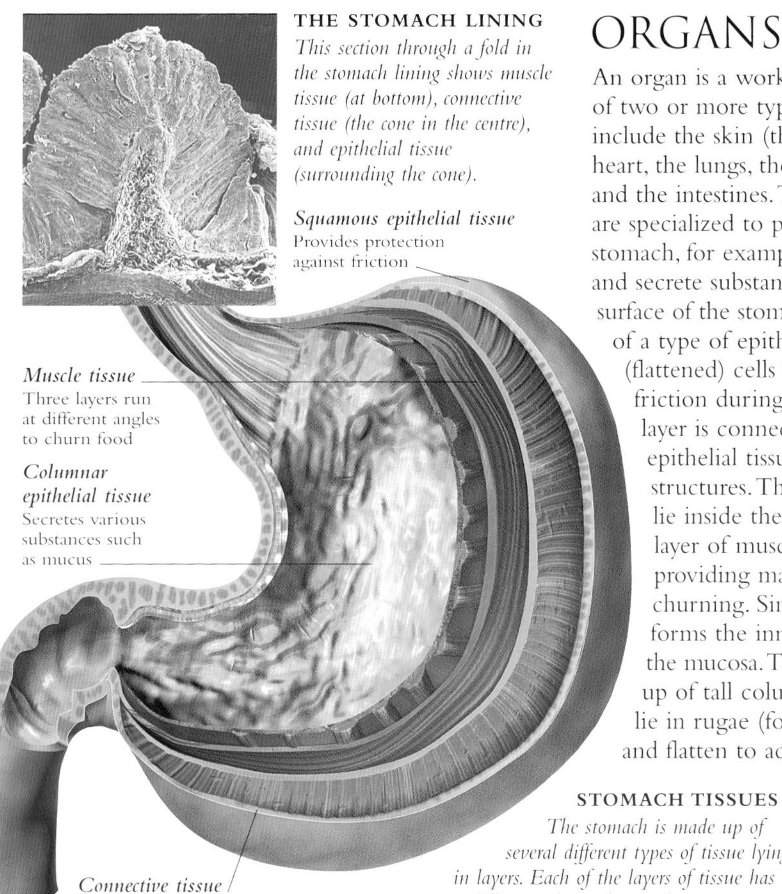

THE STOMACH LINING
This section through a fold in the stomach lining shows muscle tissue (at bottom), connective tissue (the cone in the centre), and epithelial tissue (surrounding the cone).

STOMACH TISSUES
The stomach is made up of several different types of tissue lying in layers. Each of the layers of tissue has a different role to play in the structure and function of the stomach.

ORGANS OF THE BODY

An organ is a working part of the body, consisting of two or more types of tissue. Human organs include the skin (the largest organ in the body), the heart, the lungs, the stomach, the liver, the kidneys, and the intestines. The tissues that make up organs are specialized to perform specific functions. The stomach, for example, needs to be able to churn food and secrete substances that aid digestion. The exterior surface of the stomach's outer layer, the serosa, consists of a type of epithelial tissue made up of squamous (flattened) cells that protect the stomach against friction during churning. The serosa's interior layer is connective tissue that supports the epithelial tissue and nourishes the surrounding structures. Three layers of smooth muscle tissue lie inside the connective tissue layer. Each layer of muscle tissue runs at a different angle, providing maximum directional movement for churning. Simple columnar epithelial tissue forms the inner lining of the stomach, called the mucosa. This type of epithelial tissue is made up of tall columnar cells, so that the tissue can lie in rugae (folds) when the stomach is empty and flatten to accommodate incoming food when necessary. The epithelial tissue in the lining of the stomach also houses glands that secrete enzymes and acid to break down food as well as cells that produce protective mucus.

Health

Organ transplants

Diseased or damaged organs in the body can be replaced with organs donated from another person. However, molecules in tissue vary chemically from person to person, so a "match" (a chemical agreement) must first be found to ensure that the donated organ will be accepted by the recipient body. Most matches are found between members of the same family. Commonly transplanted organs include the heart, liver, kidneys, and lungs. Tissue itself can also be transplanted.

LIVER TRANSPLANT
A donor liver is placed on a sterile cloth over the clamped opening of a patient's abdomen, ready to be transplanted.

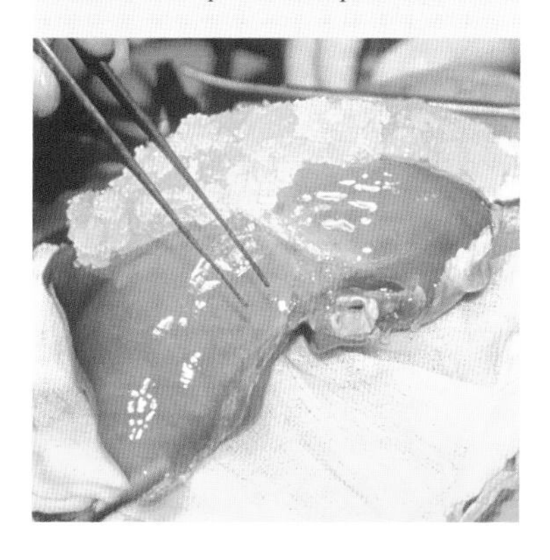

BODY SYSTEMS

Systems of the body are made up of a group of tissues and organs that work together to carry out a specific function or set of functions. For example, the musculoskeletal system consists of bone, muscle, cartilage, and tendons, which together provide support for the body and enable us to move. The main functions of each body system are listed in the table at right and their components are described on p57. Systems cannot work alone: each system is dependent on the others to function. For example, all systems in the body are reliant on the cardiovascular system to bring them nutrient- and oxygen-rich blood that provides them with the energy they require to function. The nervous system and the endocrine system are the body's control systems: they continuously monitor body activities and adjust them appropriately.

THE BODY SYSTEMS
Each of the body's systems has its own set of functions. They must not only fulfil these, but also must work together as a complementary unit to ensure the smooth functioning of the entire body.

BODY SYSTEMS AND THEIR FUNCTIONS

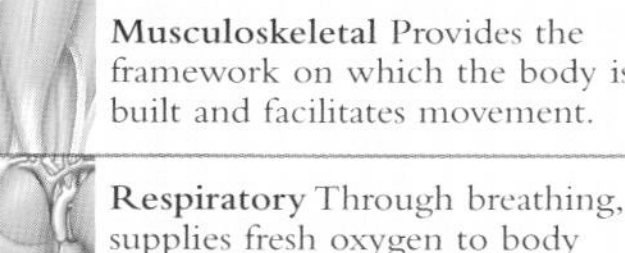

Musculoskeletal Provides the framework on which the body is built and facilitates movement.	 **Integumentary** Provides protection from the environment through the skin, hair, and nails.
Respiratory Through breathing, supplies fresh oxygen to body tissues and expels carbon dioxide.	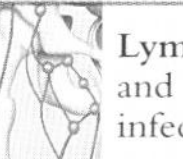 **Lymphatic and immune** Defend and protect the body from infection and some cancers.
Cardiovascular Circulates blood to deliver nutrients and oxygen to all body tissues.	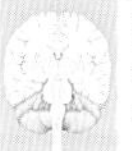 **Nervous** Senses the environment; through nerve impulses, monitors and controls body activities.
Digestive Fuels the body by food breakdown and processing of nutrients; also removes waste.	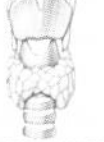 **Endocrine** Controls the body through the action of hormones secreted by glands and tissues.
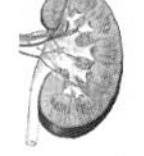 **Urinary** Forms urine to rid the body of waste and help maintain its chemical balance.	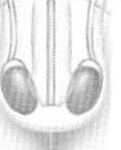 **Reproductive** Makes new bodies through the production of hormones, sperm, and eggs.

THE COMPLETE BODY

It is astonishing to think that our complex bodies are built up of simple building blocks – cells. Individual cells of similar size and shape build up into different types of tissue, each of which has a specific function. Organs, such as the stomach, are the working parts of the human machine. They are made up of two or more types of tissue that work together to carry out a particular function: for example, the stomach churns food and the ovaries produce eggs. A system is a group of tissues and organs that perform a body function, such as digestion or reproduction. Systems are dependent on each other to create a healthy, fully functioning human body.

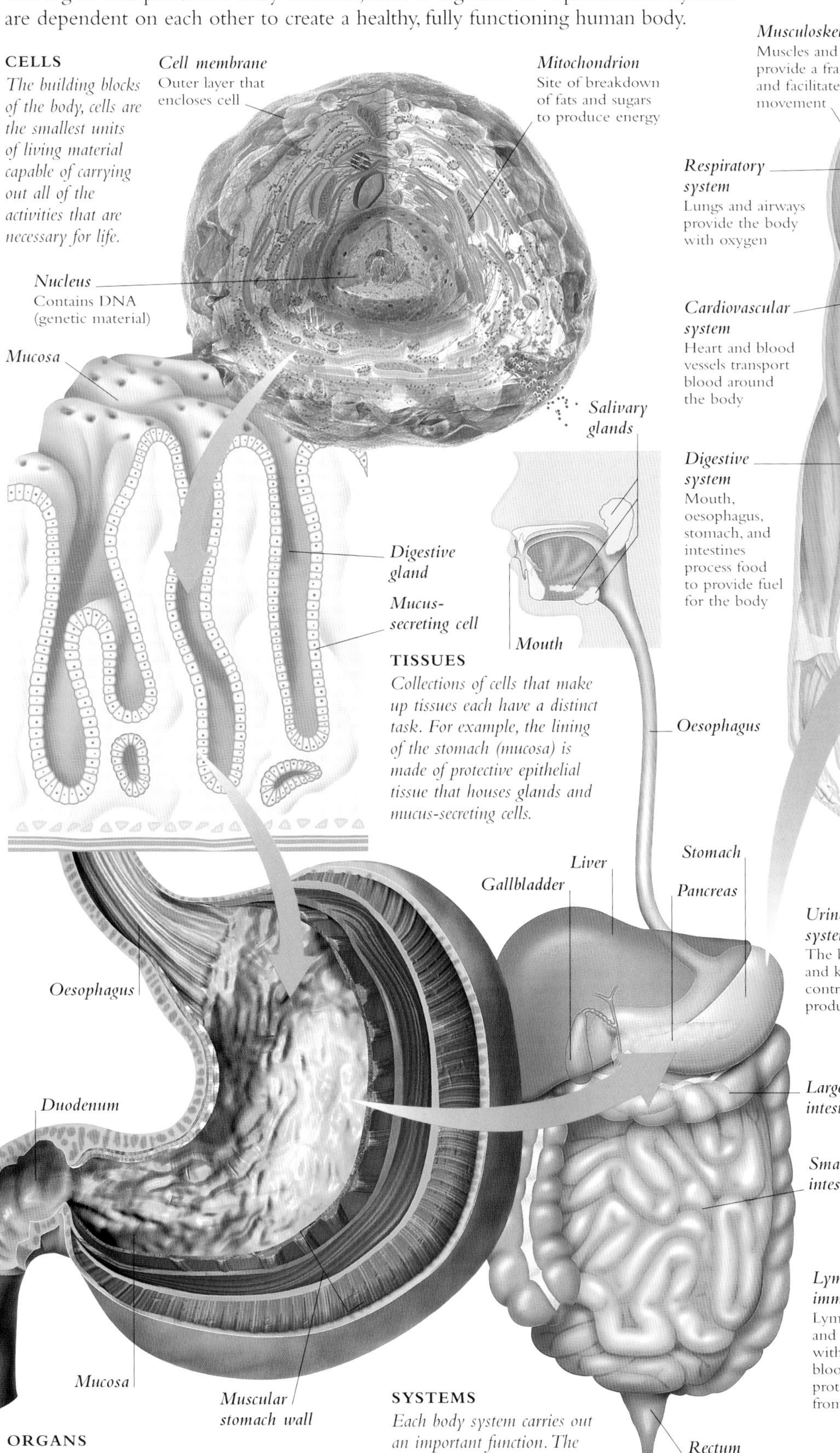

CELLS
The building blocks of the body, cells are the smallest units of living material capable of carrying out all of the activities that are necessary for life.

TISSUES
Collections of cells that make up tissues each have a distinct task. For example, the lining of the stomach (mucosa) is made of protective epithelial tissue that houses glands and mucus-secreting cells.

ORGANS
Made up of several different types of tissue, organs have key roles in each body system. The stomach's role in the digestive system is to churn, store, and partially digest food.

SYSTEMS
Each body system carries out an important function. The organs of the digestive system, assisted by accessory organs such as the liver, break down food into nutrients and process waste.

THE BODY
In a healthy body, all of the body systems work efficiently and in synchrony, so that the human body can function and reproduce.

Nervous system
Consists of the brain, spinal cord, and connecting nerves; controls all other body systems

Endocrine system
Glands, such as the thyroid, secrete hormones to regulate body functions

Musculoskeletal system
Muscles and bones provide a framework and facilitate movement

Respiratory system
Lungs and airways provide the body with oxygen

Cardiovascular system
Heart and blood vessels transport blood around the body

Digestive system
Mouth, oesophagus, stomach, and intestines process food to provide fuel for the body

Urinary system
The bladder and kidneys control urine production

Reproductive system
The testes, penis, and sperm-carrying ducts in males; ovaries, uterus, and the vagina in females are involved in the reproductive process

Integumentary system
Skin and its glands, hair, and nails protect and regulate body temperature

Lymphatic and immune system
Lymph vessels and nodes, work with white blood cells to protect the body from disease

MOVING THE BODY

Muscles provide the pulling power that bends the joints, lifts the legs, and propels the body forwards.

SUPPORT AND MOVEMENT

Humans are vertebrates, creatures that have a bony inner scaffolding with a central spine. Layer upon layer of muscles clothe these bones, forming solid flesh beneath the skin. Where bones meet, joints of various types allow this remarkable construction to move with great versatility. Although we are flexible, our sturdy core structure and limbs keep us balanced in our unique two-legged stance. Highly developed brain–muscle coordination enables us to fine-tune movement to a degree no other animal achieves.

Bones, joints, and muscles are all interdependent. If the skeleton were not provided with muscles it would collapse in a loose tangle of bones, like a puppet without its strings. Without joints, the body would for the most part be rigid and immobile. However, a few joints, such as those between the bones of the lower leg, provide stability, rather than flexibility.

A BONY FRAMEWORK

Our long column of linked spinal bones (vertebrae) curves gently backwards and forwards over the centre of gravity. This arrangement provides the spine with greater resilience to shocks and gives stability to our upright posture. All other bones in the body link directly or indirectly to the spine, and each bone is designed for a particular purpose. They are the anchor points for muscles and the core around which soft tissues are wrapped. Some bones, such as those of the pelvis, are weight bearing, others have a protective function, girdling vital organs such as the heart and lungs and encasing the vulnerable brain.

Bones are long, short, flat, round, and various other shapes in between. Adult skeletons are readily distinguishable as either male or female by their size and small differences in their shape. In general, the bones of a male are slightly larger and heavier than those of a female because the muscles they have to support are heavier.

Bones usually stay strong and resilient well into old age, despite the stresses we often impose upon them. This is because they are able to renew themselves. Each bone contains living cells that are in a state of slow but continuous action, breaking down and renewing tissue. This process continues throughout life, keeping bones in good repair. If a bone breaks it is able to mend itself so successfully that within a few months it regains full strength, with few traces of the damage.

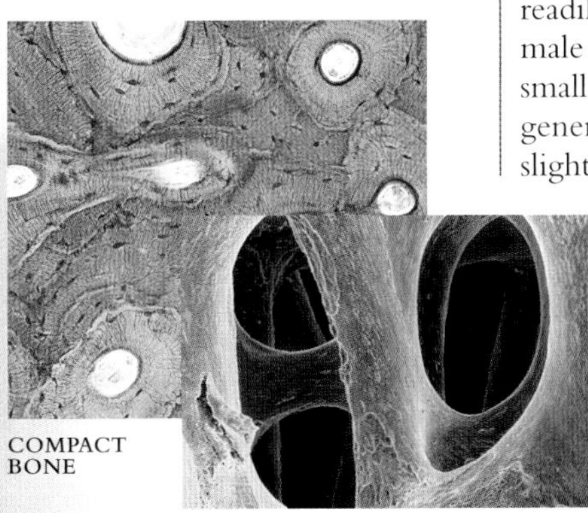

BONE STRUCTURE
A long bone, such as in the arm, has a dense outer layer called compact bone, arranged in concentric rings. The inner layer, known as spongy bone, is an open network of rigid struts (above).

FLEXIBILITY

We can perform an amazing variety of physical feats because of the many ways in which our joints move. A joint is a highly elaborate piece of engineering. It involves bones that swivel, tilt, and slide; muscles that pull; and tough tendons and ligaments that hold everything securely together. From babyhood, one stage at a time, we find out how these joints make our bodies move. We learn to walk, run, and kick a ball. Later, we progress to movements requiring more sophisticated control and coordination, such as playing the piano or dancing. A few exceptional humans, such as athletes or gymnasts, train their bodies to perform movements that go beyond normal limits of flexibility and endurance.

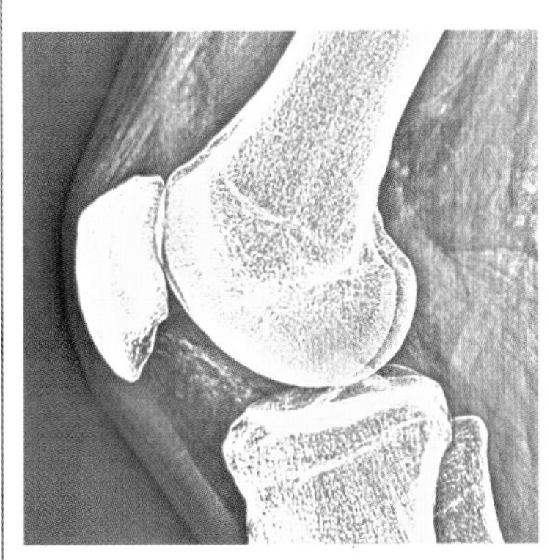

MOVABLE JOINT
Flexible joints, such as the knee shown in this scan, allow us to move limbs and body in many different directions.

MUSCLE ACTION

The muscles that make us move are known as skeletal muscles. They vary greatly in size, ranging from the tiny muscles that move our eyes around in their sockets to the large muscles of the back and upper legs. Muscle tissue is dense and heavy. Together, the skeletal muscles make up about 40 per cent of our total body weight.

Muscles are attached to bones by cord-like tendons and make bones move by exerting a pull. Typically, they work in pairs, producing opposing movements. Control of every muscle movement that we make comes from the brain. Whether it is subconscious or voluntary, each movement is the result of a nerve signal reaching a particular muscle This signal stimulates the muscle fibres, which contract rapidly. However, much of the time muscle action is barely noticeable. For example, when we stand at a bus stop we may not be aware of any physical effort, but our muscles are at work. They automatically make constant tiny adjustments, correcting their tension to hold the spine erect, the head poised without wobbling on the neck, and body weight balanced centrally over the feet.

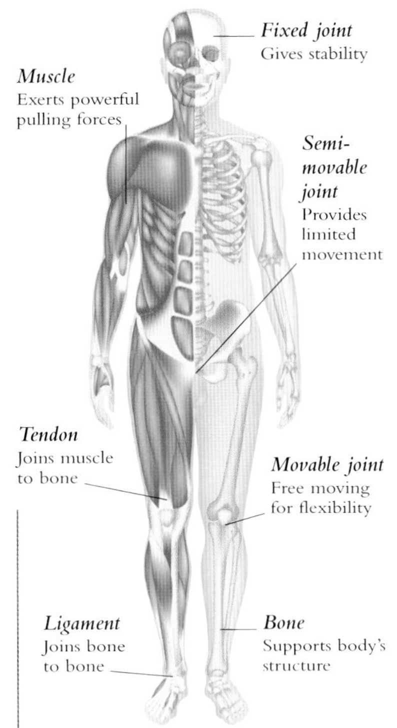

MOVING FRAMEWORK
Body movement and stability depend on precisely controlled interaction between bones, muscles and joints.

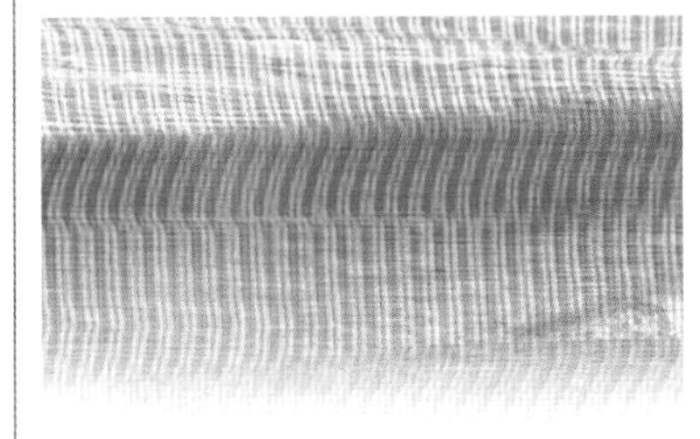

MUSCLE FIBRES
When seen highly magnified, the long fibres of skeletal muscles have a striped appearance. These muscles are attached to bones and enable us to move.

History

Focus on movement

Human movement was revealed in detail for the first time through the work of a pioneering English photographer called Eadweard Muybridge (1830–1904). Following his revolutionary studies of animal locomotion, which excited worldwide attention, Muybridge turned to human subjects. In the 1880s, he took numerous sequence photographs, such as the somersaulting man shown right, that captured nuances of movement that are imperceptible to the onlooker.

THE SKELETON

The adult skeleton is made up of 206 individual bones. These come in all shapes and sizes from the tiny bones of the inner ear, measuring just a few millimetres, to the massive bones of the pelvis. Each bone links with others to form a sturdy, flexible framework. This structure is designed to perform many functions. The whole skeleton supports the soft tissues of the body and gives us shape. Specific groups of bones, powered by their attached muscles, provide the leverage that allows us to make a wide range of coordinated movements. As well as giving us shape and mobility, the skeleton provides protection for vital internal organs.

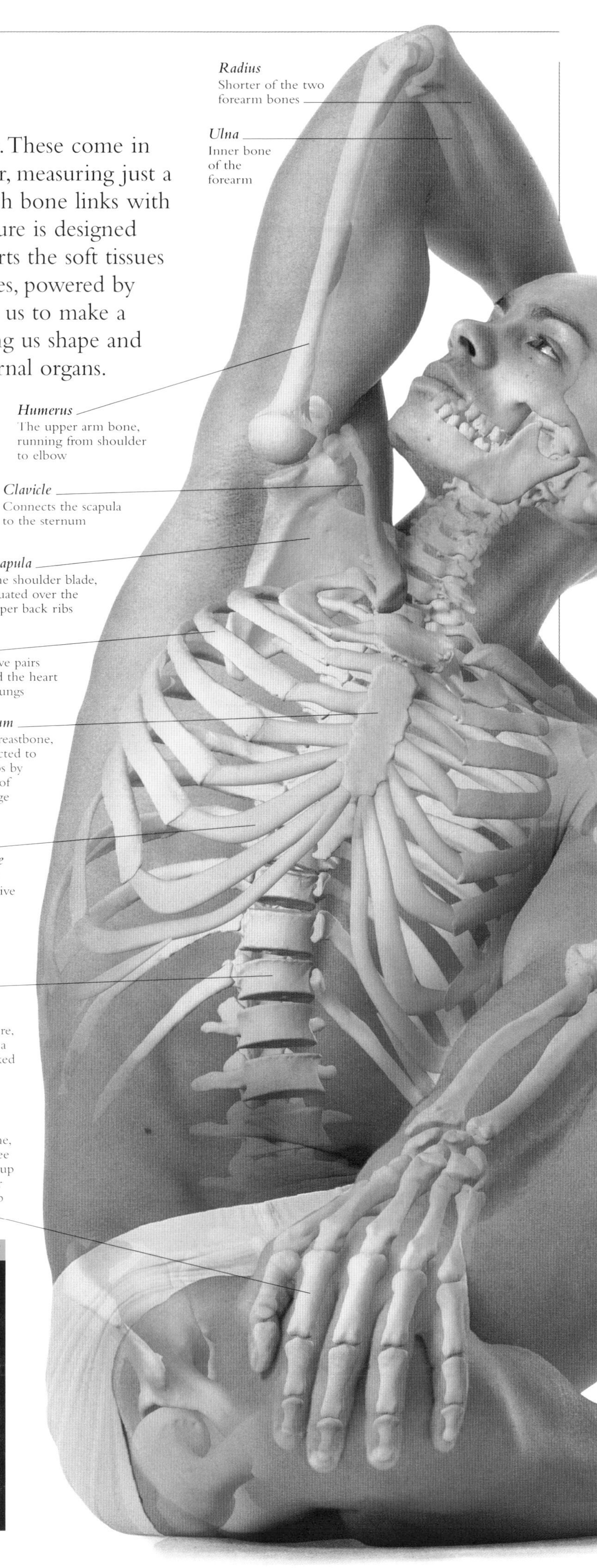

SKELETAL ROLES

Our bones are arranged symmetrically on either side of the body and are organized in two main divisions. These divisions, the axial and appendicular skeletons, have different roles. The axial skeleton comprises the bones of the central part of the body. These bones include the skull, the spinal column, the ribs, and the sternum (breastbone) – 80 bones in all. Such bones have a predominantly protective role, surrounding some of the body's most vulnerable and vital parts. The bony helmet of the skull encases the brain, the bones of the spinal column contain the spinal cord, and the ribs form a cage around the heart and the lungs. Attached to the axial skeleton are the 126 bones of the appendicular skeleton, the main role of which is to provide the body with mobility. Bones in this division, which include those of the arms and legs, have many joints. Also part of the appendicular skeleton are the shoulder blades and the pelvis, which are the linking structures between the limbs and the core of the body.

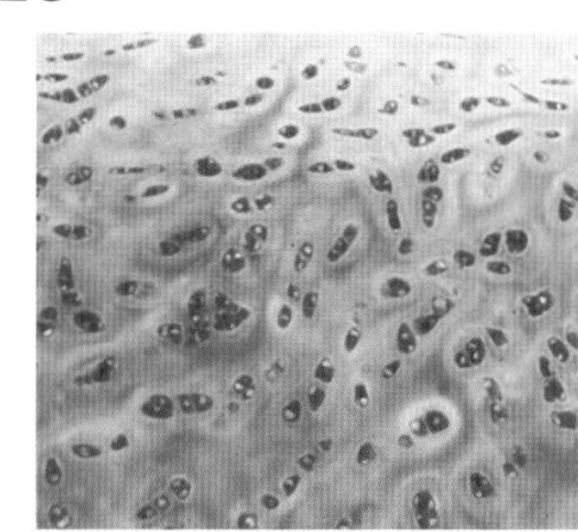

COSTAL CARTILAGE
Pliable costal cartilage, a type of connective tissue, joins the ten upper pairs of ribs to the breastbone. The dark areas here are developing cartilage cells, surrounded by fibrous protein.

SKELETAL DIVISIONS
The appendicular skeleton (blue) comprises the arms and legs and the bony girdles that link the limbs to the axial skeleton, the central frame of the body.

History

Early X-rays

The first-ever X-rays of the human skeleton were taken in 1895. The technique was accidentally discovered by Wilhelm Roentgen, a German physicist. He observed that certain electromagnetic rays passed through soft body tissues but were blocked by dense bone. When Roentgen placed his wife's hand in front of the rays, the bones of her fingers, and her ring (shown here), appeared as shadow images on a photographic plate.

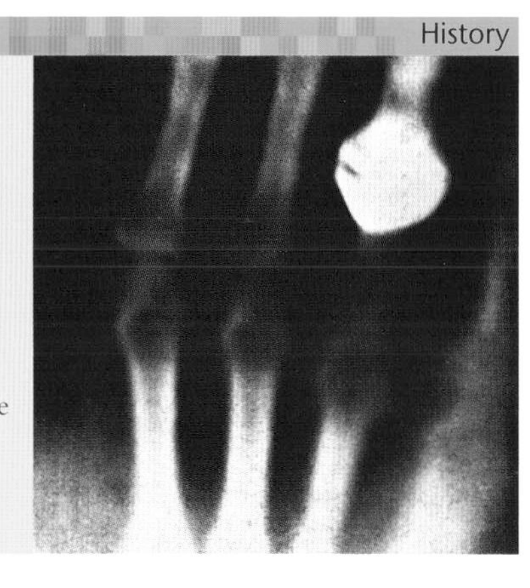

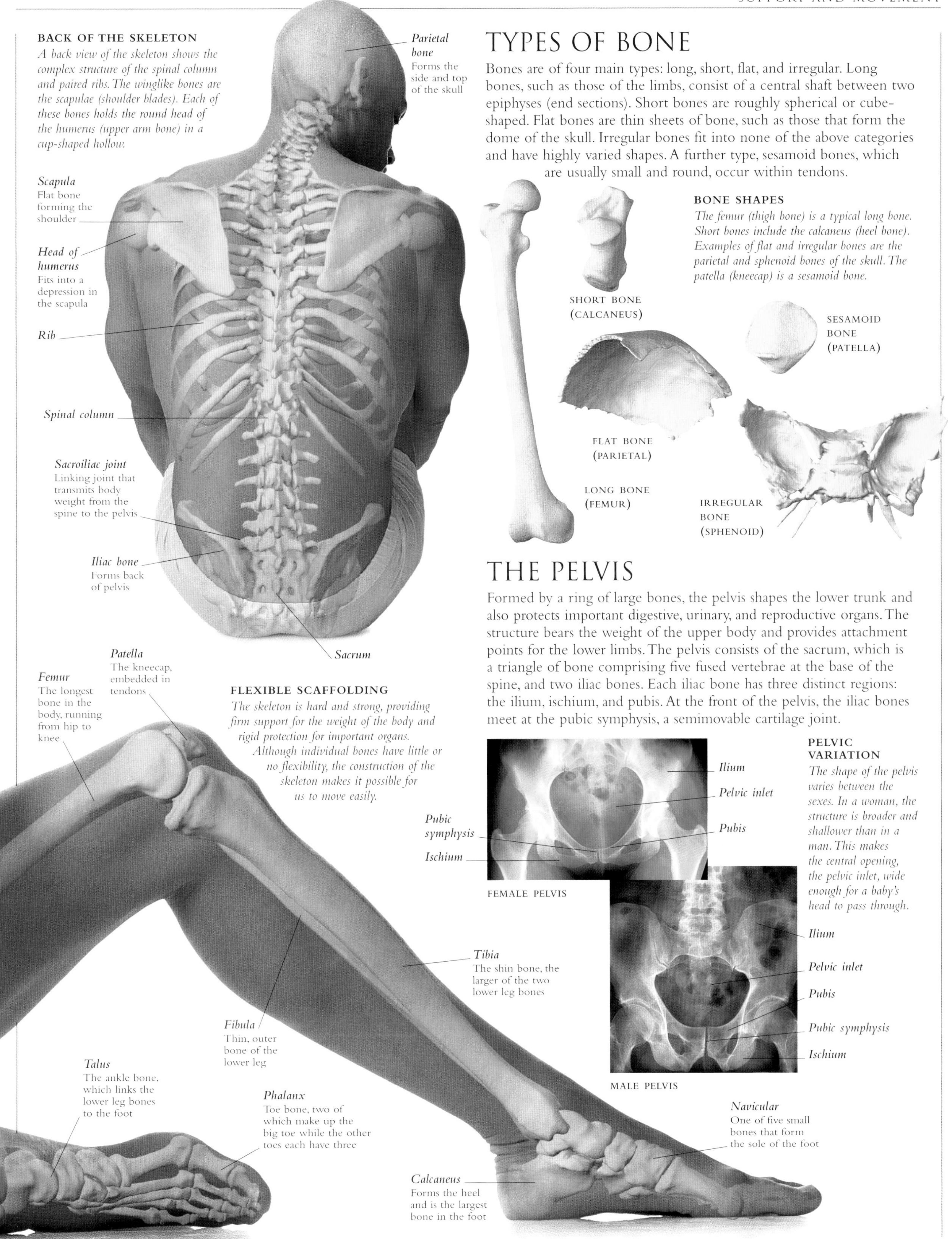

BACK OF THE SKELETON
A back view of the skeleton shows the complex structure of the spinal column and paired ribs. The winglike bones are the scapulae (shoulder blades). Each of these bones holds the round head of the humerus (upper arm bone) in a cup-shaped hollow.

TYPES OF BONE

Bones are of four main types: long, short, flat, and irregular. Long bones, such as those of the limbs, consist of a central shaft between two epiphyses (end sections). Short bones are roughly spherical or cube-shaped. Flat bones are thin sheets of bone, such as those that form the dome of the skull. Irregular bones fit into none of the above categories and have highly varied shapes. A further type, sesamoid bones, which are usually small and round, occur within tendons.

BONE SHAPES
The femur (thigh bone) is a typical long bone. Short bones include the calcaneus (heel bone). Examples of flat and irregular bones are the parietal and sphenoid bones of the skull. The patella (kneecap) is a sesamoid bone.

THE PELVIS

Formed by a ring of large bones, the pelvis shapes the lower trunk and also protects important digestive, urinary, and reproductive organs. The structure bears the weight of the upper body and provides attachment points for the lower limbs. The pelvis consists of the sacrum, which is a triangle of bone comprising five fused vertebrae at the base of the spine, and two iliac bones. Each iliac bone has three distinct regions: the ilium, ischium, and pubis. At the front of the pelvis, the iliac bones meet at the pubic symphysis, a semimovable cartilage joint.

PELVIC VARIATION
The shape of the pelvis varies between the sexes. In a woman, the structure is broader and shallower than in a man. This makes the central opening, the pelvic inlet, wide enough for a baby's head to pass through.

FLEXIBLE SCAFFOLDING
The skeleton is hard and strong, providing firm support for the weight of the body and rigid protection for important organs. Although individual bones have little or no flexibility, the construction of the skeleton makes it possible for us to move easily.

THE SKULL

The skull's most important role is to enclose and protect the brain, but it also gives shape to the head and face, and houses the sense organs, forming sockets for the eyes and cavities for the sinuses and nasal passages. Two separate sets of bones form the skull. The eight bones surrounding the brain, known as the cranial vault, are mostly large curved plates of bone. A further 14 bones, of various shapes and sizes, make up the skeleton of the face. During childhood, the bones of the skull yield to facilitate growth, but by adulthood, all of the bones, except the lower jaw (*see below*), are fused together.

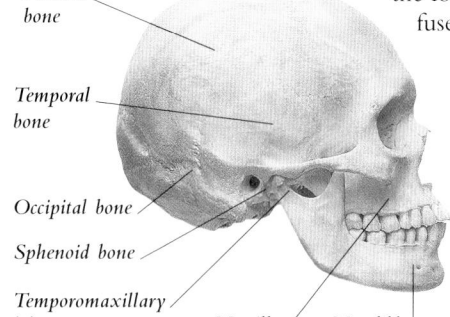

Parietal bone
Temporal bone
Occipital bone
Sphenoid bone
Temporomaxillary joint
Maxilla
Mandible

SKULL MOVEMENT
The U-shaped mandible (lower jaw) is the only movable bone in the skull. It slots into the cranial vault, forming two hinges at ear level, which allow the jaw bone to move up and down, opening and closing the mouth. It also houses the bottom set of teeth.

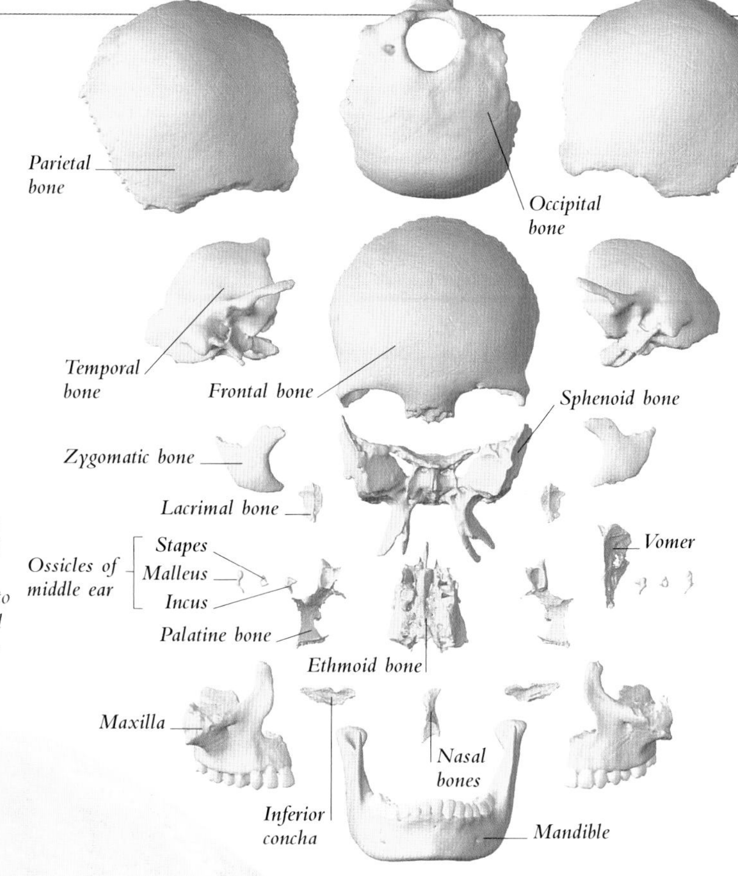

BONES OF THE SKULL
The bones of the skull vary widely in size and shape – from the large, smooth, curved parietal and occipital bones to the tiny, intricate, jagged ossicle bones of the middle ear. The ossicles of the middle ear are the smallest bones in the body.

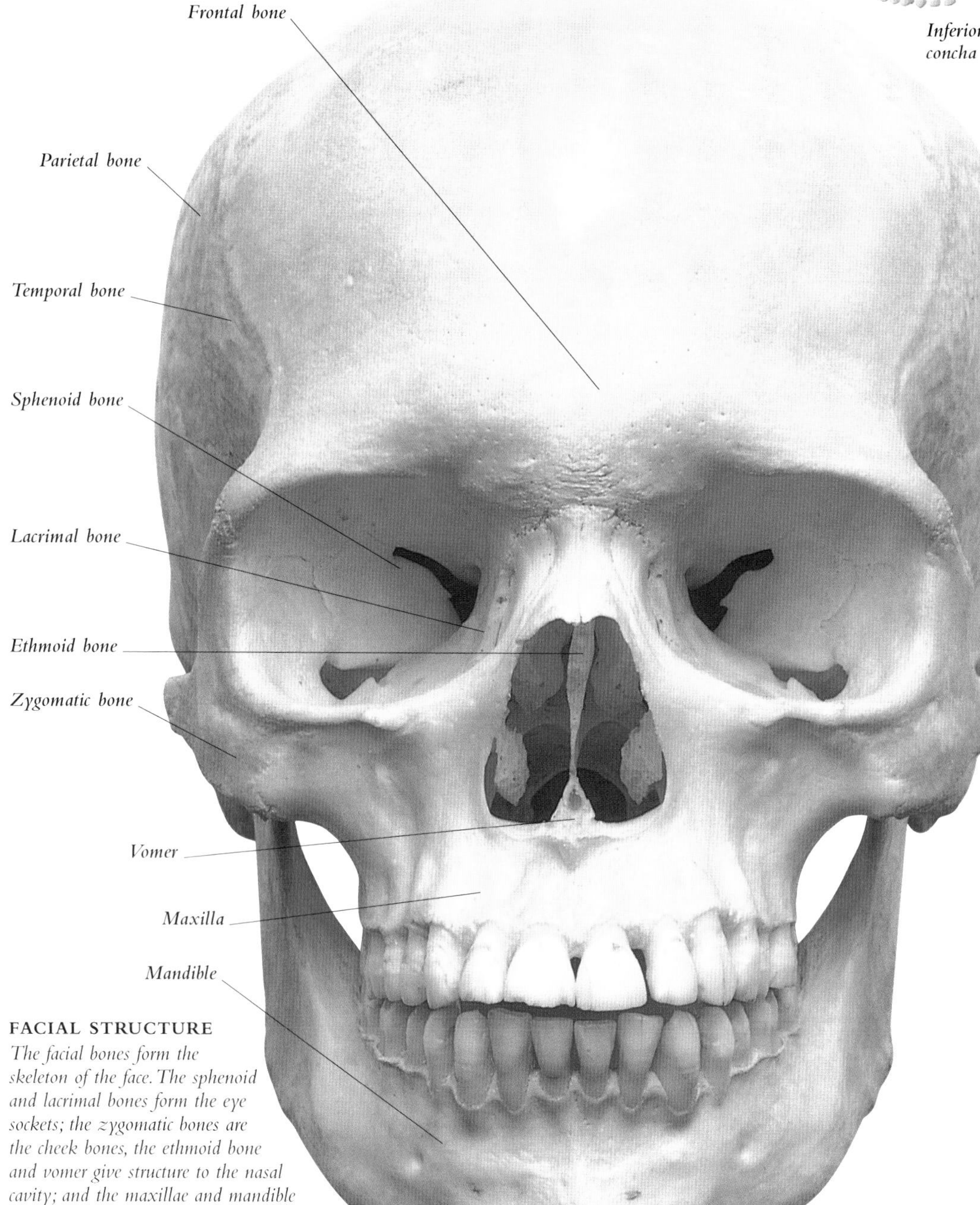

FACIAL STRUCTURE
The facial bones form the skeleton of the face. The sphenoid and lacrimal bones form the eye sockets; the zygomatic bones are the cheek bones, the ethmoid bone and vomer give structure to the nasal cavity; and the maxillae and mandible contain sockets for the teeth.

History

Trepanning

The practice of trepanning consists of boring a hole in the skull of a living patient and removing a piece of bone, leaving the membrane surrounding the brain exposed. In Neolithic times, the practice was widespread, although the reasons for it are not known. It may have been performed to allow spirits to enter or exit the brain; to relieve headaches, infections, and convulsions, or even provide a cure for insanity; or to acquire rondelles (discs of bone) for charms or amulets. Incredibly, the practice of trepanning is still performed today in parts of Africa and South America.

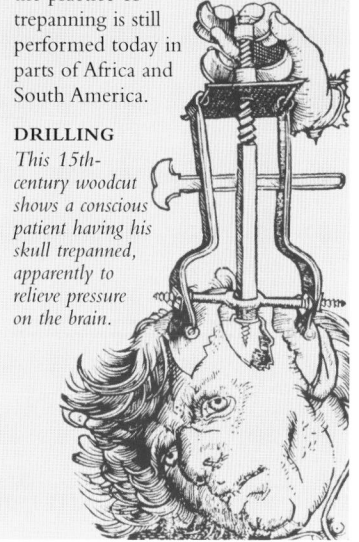

DRILLING
This 15th-century woodcut shows a conscious patient having his skull trepanned, apparently to relieve pressure on the brain.

BODY

THE SPINE

The spine is made up of 33 bones called vertebrae, which are linked by a series of flexible facet joints. There are three main types of vertebra: cervical, which support the head and neck; thoracic, which secure the ribs; and lumbar, which support a large proportion of the weight of the upper body. The triangular-shaped sacrum and the tail-like coccyx consist of a number of vertebrae fused together. These bones sit at the base of the spine, providing a solid foundation for the spinal column.

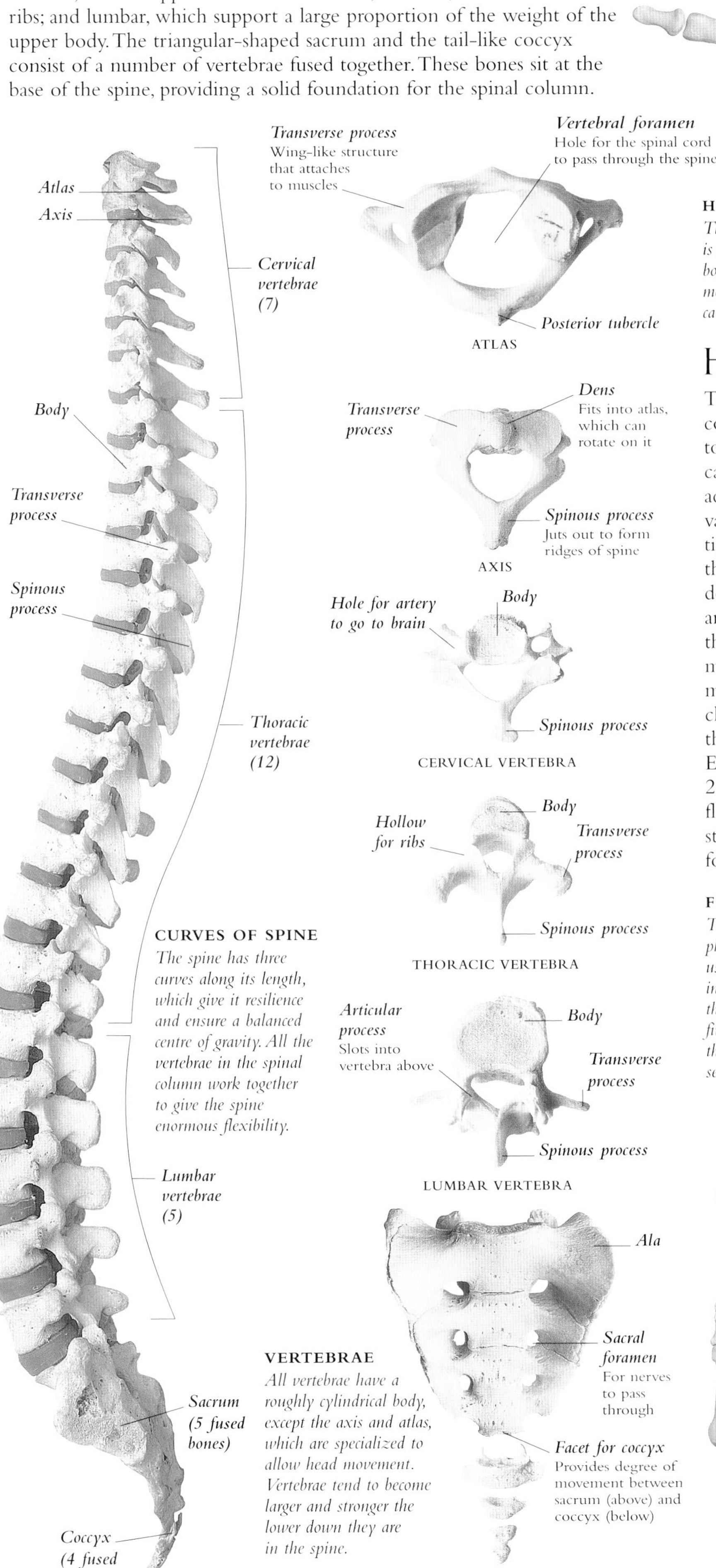

CURVES OF SPINE

The spine has three curves along its length, which give it resilience and ensure a balanced centre of gravity. All the vertebrae in the spinal column work together to give the spine enormous flexibility.

VERTEBRAE

All vertebrae have a roughly cylindrical body, except the axis and atlas, which are specialized to allow head movement. Vertebrae tend to become larger and stronger the lower down they are in the spine.

Hamate
Pisiform
Lunate
Scaphoid
Triquetrum
Capitate
Trapeziod
Trapezium
Metacarpals
Phalanges
Carpals

HAND BONES

The hand (the back of which is seen here) is made up of three different types of bone: 14 phalanges (finger bones), five metacarpals (palm bones), and eight carpals (wrist bones).

HAND AND FOOT BONES

The hands and feet are similar in their skeletal make-up; both comprise an interlinking arrangement of small bones, which together form a fan-like structure. The hand is a versatile tool, capable of delicate manipulation as well as powerful gripping actions. The arrangement of its 27 small bones allows for a wide variety of movements. In particular, it is the ability to bring the tip of the thumb and the fingers together (having opposable thumbs) that gives human hands their unique dexterity. The feet and toes support and propel the entire weight of the body while it is on the move and also help to maintain balance during changes of position while the body is stationary. Each foot has a total of 26 bones, forming a flexible platform of strength and support for the entire body.

FOOT BONES

The foot has 14 phalanges (toe bones), usually shorter than those in the hand. The rest of the foot is composed of five metatarsals (forming the sole of the foot) and seven tarsals (ankle bones).

Medial cuneiform
Intermediate cuneiform
Navicular
Lateral cuneiform

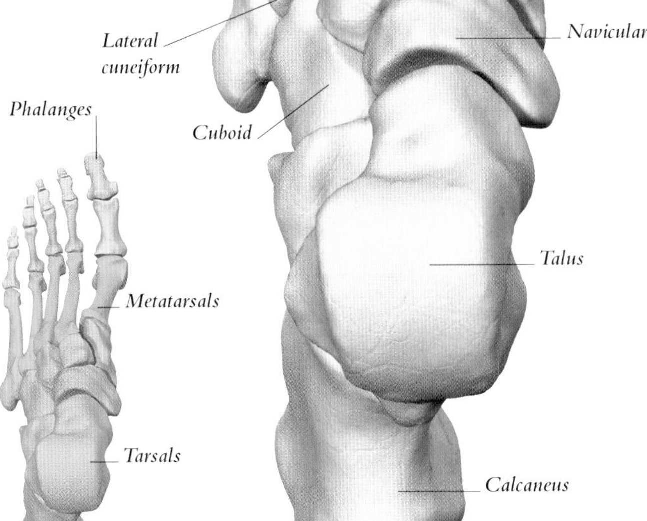

BODY

BONE STRUCTURE AND GROWTH

Although bone looks solid and is extremely sturdy, it is surprisingly light; it also has a slight flexibility that gives some protection against shocks and jarring. These qualities are due to bone's remarkable structure and its fabric of elastic protein fibres. Bone is not inert but is constantly being remodelled, even after we reach full adult growth. One of bone's vital activities is the manufacture of most of our blood cells. Bone also serves as a reservoir for various minerals, such as calcium and phosphorus.

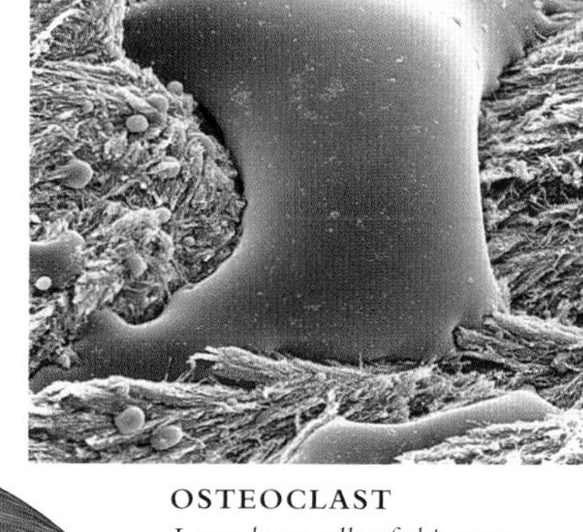

OSTEOCLAST
Large bone cells of this type constantly break down bone tissue so that it can be replaced with newer tissue.

A LIVING TISSUE

Bone is made up of living cells, as well as protein, mineral salts, and water. The cells of bones are of two main types: osteoblasts, which build new bone, and osteoclasts, which break it down. This process, which is lifelong, keeps our bones in a constant state of renewal and helps to minimize wear and tear until well into old age. Regardless of shape or size, each bone has an outer layer of compact bone, which is a dense heavy tissue, and an inner layer of light, spongy bone, which comprises an open network of interconnecting struts. Compact bone is made up of rod-shaped units called osteons. An osteon consists of concentric layers of hard tissue around a central channel containing blood vessels and nerves. Between the layers are tiny spaces housing bone cells and fluid that provides the cells with nutrients. In spongy bone, the gaps between the struts are filled with bone marrow, which in some bones is the site of blood cell manufacture (*see* p83).

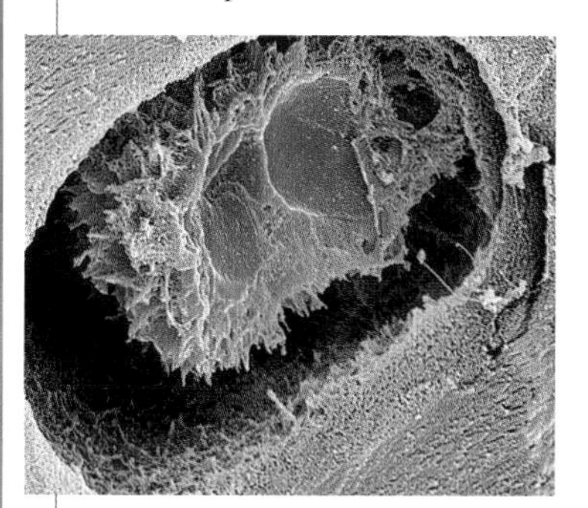

OSTEOBLAST
In this image, an osteoblast has become trapped inside a cavity in compact bone. These bone-producing cells maintain bone strength.

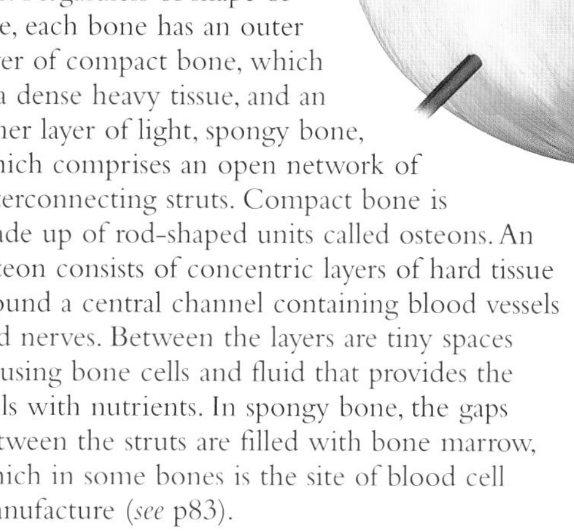

COMPACT BONE
Dense, heavy outer layer of bone that is one of the toughest materials in the body
The units of compact bone, osteons, are arranged in rings. The white areas are channels for blood vessels and nerves. The dark spots are spaces containing bone cells.

BONE DEVELOPMENT AND GROWTH

When the skeleton first develops in a young fetus, it is made entirely of tough, springy cartilage. By the time a baby is born, however, much of the cartilage has hardened into bone tissue. The conversion of cartilage into bone, ossification, begins at sites called primary ossification centres in the shafts of long bones. In a newborn baby, the bone shafts are completely hardened, but the bone ends, known as the epiphyses, are still cartilage. Within these cartilage ends, hard bone gradually develops from secondary ossification sites. Between the shaft and the ends of a bone is a zone, the growth plate, which produces more cartilage to elongate bones. The process of ossification continues until the age of about 18 years. By adulthood, the processes of growth and ossification are complete and both the shaft and the ends of the bone have become continuous bone.

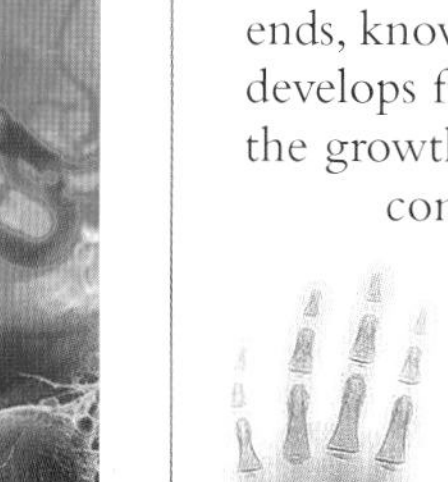

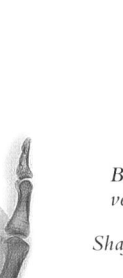

HAND GROWTH
In a child aged 2, ossified hand bone shafts are opaque on an X-ray. The apparent gaps between the joints are cartilage, which looks transparent. By age 18, all bone is ossified.

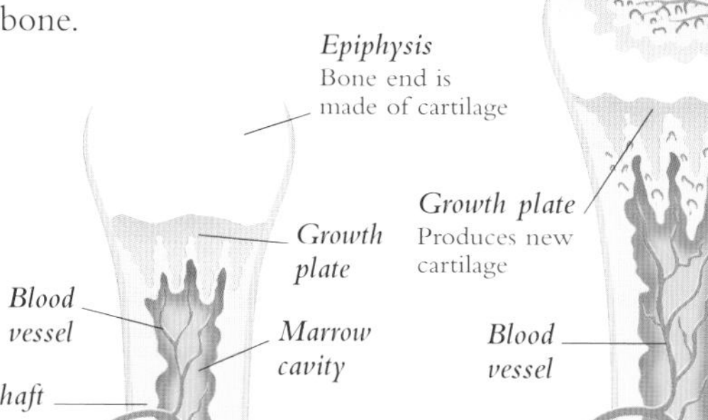

Marrow cavity

LONG BONE OF A NEWBORN BABY
The shaft is mostly bone, while the ends are made of cartilage that will gradually harden.

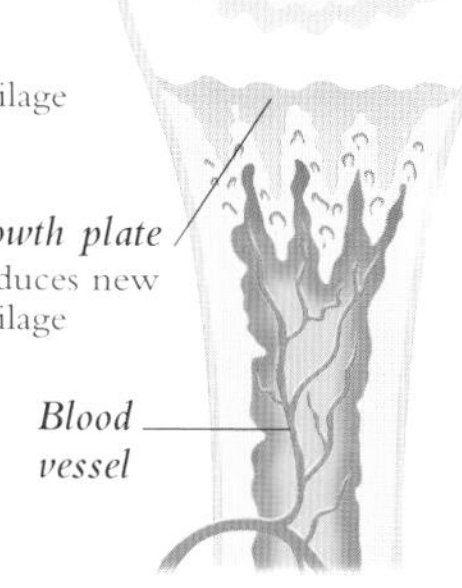

LONG BONE OF A CHILD
Bone is forming in secondary ossification centres in the ends. A growth plate near each end of the bone produces new cartilage.

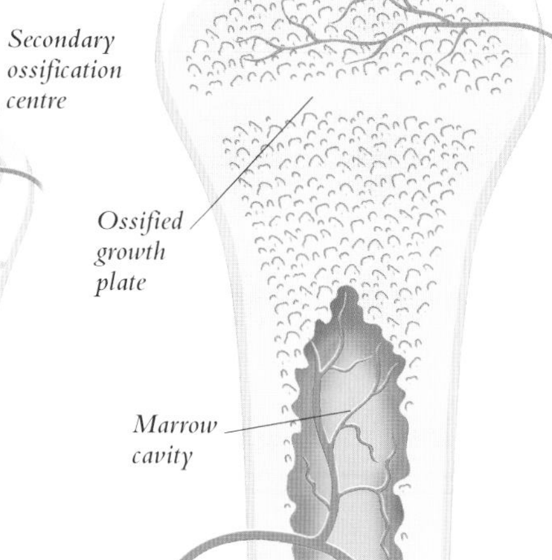

Marrow cavity

LONG BONE OF AN ADULT
Growth is complete by age 18. The shafts, growth plates, and bone ends have all ossified and fused into continuous bone.

BODY

BONE REPAIR

Bone has an astonishing ability to mend itself after a fracture. The healing process begins within minutes of the break, when the blood clotting process is activated (*see* Repairing injuries, p105). Bone cells rapidly start building up a mass of new spongy tissue, called the callus, around the site of the damage. This tissue gradually becomes dense compact bone. A fracture in a long bone, such as in the leg or arm, normally takes about 6 weeks to heal in an adult. In children, the process is usually quicker. For some months after healing, a swelling remains over the site of the fracture. The thickened area is gradually whittled by the action of the osteoclasts, the cells that break down bone tissue, and the bone eventually regains its normal shape. Broken bones need to be returned to, and maintained in, their correct position in order for the ends to rejoin properly. For this reason, healing bones may need to be immobilized in a plaster or resin cast until the healing process is complete. If the fracture is severe, the broken ends may be pinned together with metal nails or plates.

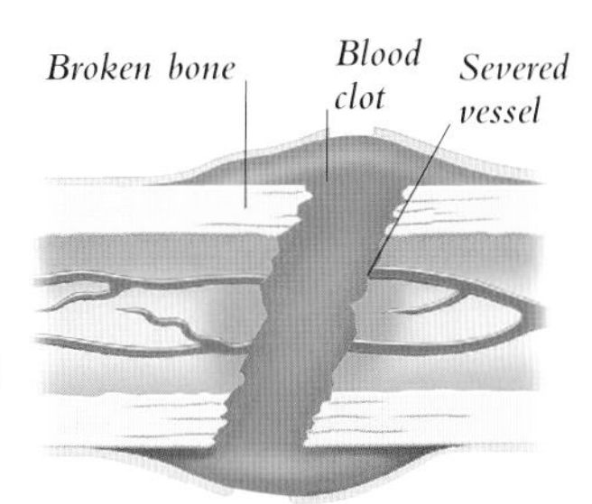

IMMEDIATE RESPONSE
The first stage of repair begins almost at once with the formation of a blood clot. This seals off leaking blood vessels within the bone.

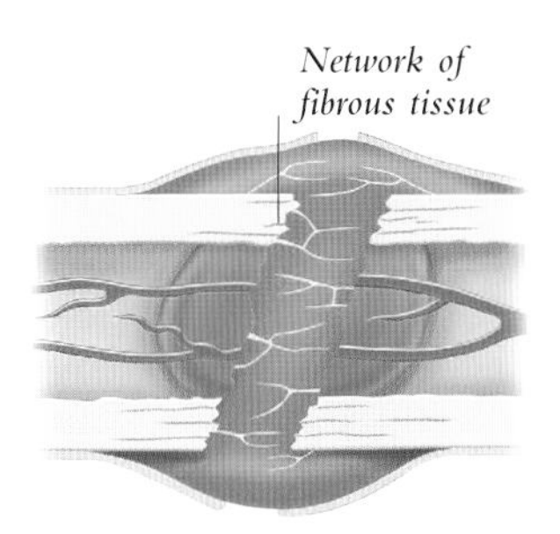

AFTER SEVERAL DAYS
A mesh of fibrous tissue forms, reaching across the gap between the broken bone ends and gradually replacing the blood clot.

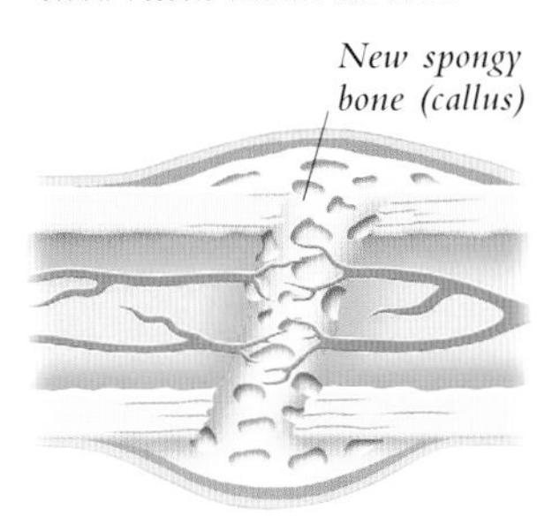

AFTER 1–2 WEEKS
New, soft spongy bone, called callus, develops on the framework of the fibrous tissue, filling in the gap and eventually joining the bone ends.

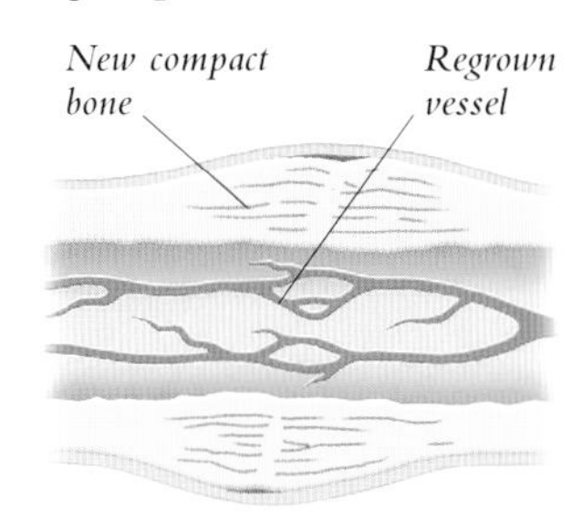

AFTER 2–3 MONTHS
Repair is almost complete. Dense compact bone replaces the callus and blood vessels have regrown. The bulge will slowly shrink away.

Osteoporosis

As people get older the rate at which their bone tissue is renewed slows down. By the age of 70 most people's skeletons have become about a third thinner and lighter than they were at age 40. This loss of bone density, called osteoporosis, makes bones more fragile and likely to break. The condition is linked to declining levels of sex hormones, and post-menopausal women are usually the most severely affected.

FRAGILE BONE
The spongy bone tissue shown below is affected by osteoporosis. Its network of struts has become porous and brittle.

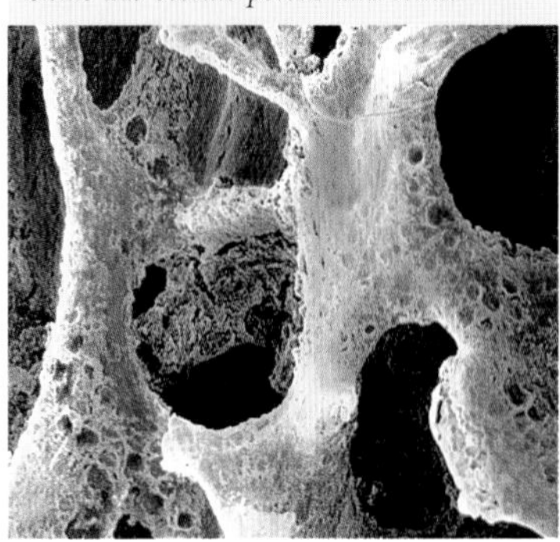

Osteon
Unit of compact bone comprising concentric layers of bone tissue

Bone marrow
The soft, fatty substance that fills the central cavities in bones and produces blood cells.

Vein

SPONGY BONE
The lightweight honeycomb structure of spongy bone, as seen in this image (left), prevents the skeleton from being excessively heavy.

Epiphysis
Forms each end of a long bone

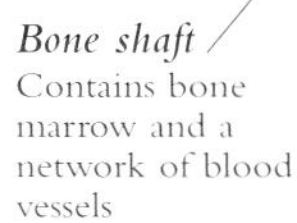

Bone shaft
Contains bone marrow and a network of blood vessels

Artery

Spongy bone
Open network of bony struts forming the bone's inner layer

BONE MARROW
The spaces between the struts of spongy bone are seen here packed with bone marrow.

STRUCTURE OF A LONG BONE
A long bone such as the femur has a central canal filled with soft bone marrow and blood vessels. The canal is surrounded by a layer of spongy bone, which is wrapped around with a layer of tough compact bone. Covering the bone's outer surface is a thin membrane, the periosteum.

Fact

Ancient bones

Because bones are so hard, they can remain undecayed for hundreds of years after death. Over immense periods of time, bones can fossilize, their tissues being replaced with even harder minerals. Fossil bones often retain their shape so well that they are immediately recognizable. This 12,000-year-old skeleton of a Cro-Magnon human enables palaeontologists to reconstruct the appearance of one of our early ancestors.

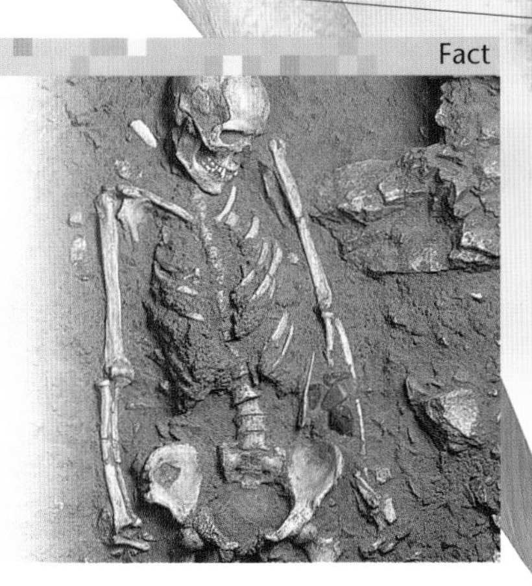

JOINTS

The point at which two bones meet is called a joint, also known as an articulation. Joints are classified by their structure or by the way in which they move. In the freely movable synovial joints, the surfaces that are in contact, called the articular surfaces, slide over each other easily. Semimovable joints, such as those in the spine, are more firmly linked and provide greater stability but less flexibility. Some joints, such as those of the skull, do not have any mobility at all.

FREELY MOVABLE JOINTS

Most of the joints in the body are freely movable synovial joints. These joints are lubricated by synovial fluid secreted by the joint lining, thus enabling articular surfaces to move with minimal friction. Pivot and hinge joints move in only one plane (from side to side, or up and down, for example), and ellipsoidal joints are able to move in two planes at right angles to each other. Most joints can move in more than two planes, which allows for a wide range of movement. The shoulder, a ball-and-socket joint, is one of the most mobile and most complex joints in the body. It moves up and down, forwards and backwards, and can even rotate, allowing the arm to move in a complete circle.

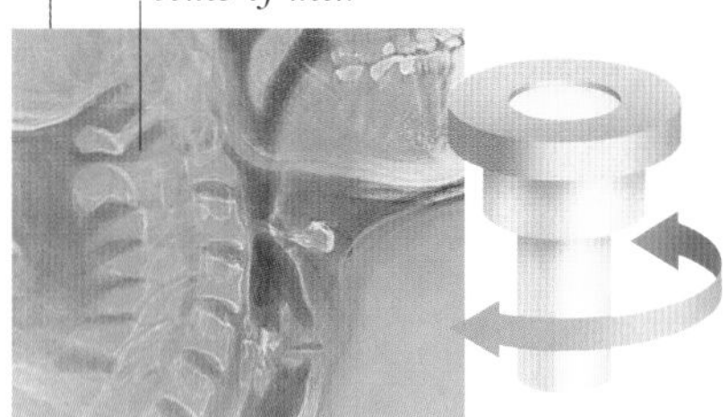

PIVOT JOINT
One bone rotates within a collar formed by another. The pivot joint between the upper bones of the neck allows the head to turn.

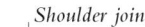

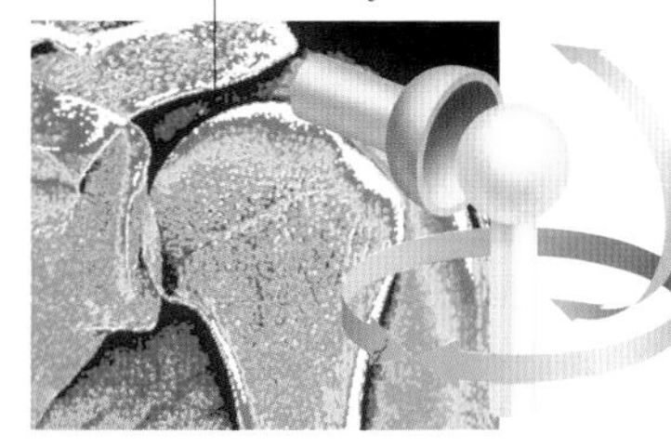

BALL-AND-SOCKET JOINT
The ball-shaped end of one bone fits into a cup-shaped cavity in another, allowing a range of movement, such as in the shoulder.

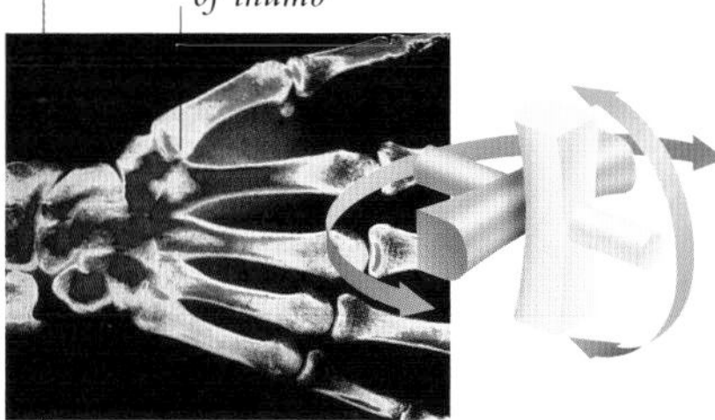

SADDLE JOINT
Saddle-shaped bone ends that meet at right angles can rotate and move back and forth. Saddle joints are at the base of the thumbs.

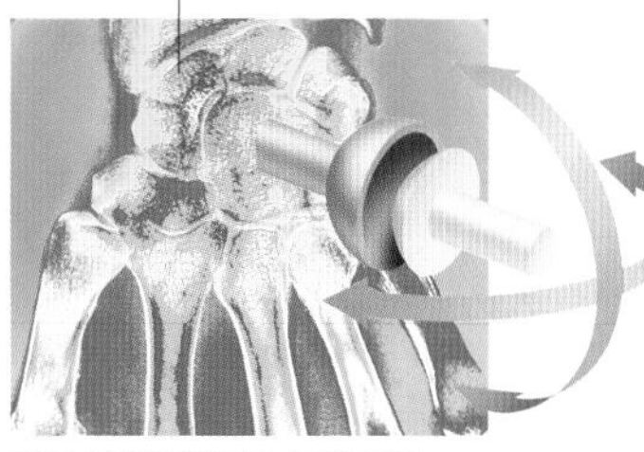

ELLIPSOIDAL JOINT
The oval end of one bone fits into the cup of another, allowing varied movement, but little rotation, such as at the wrist.

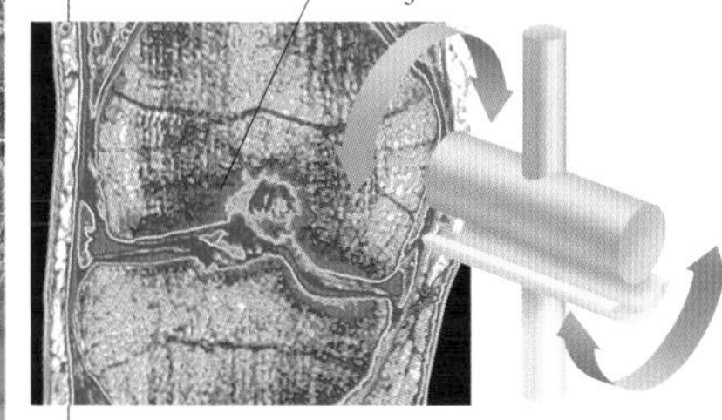

HINGE JOINT
The cylindrical surface of one bone fits into the groove of another, allowing bending and straightening, for example at the knee.

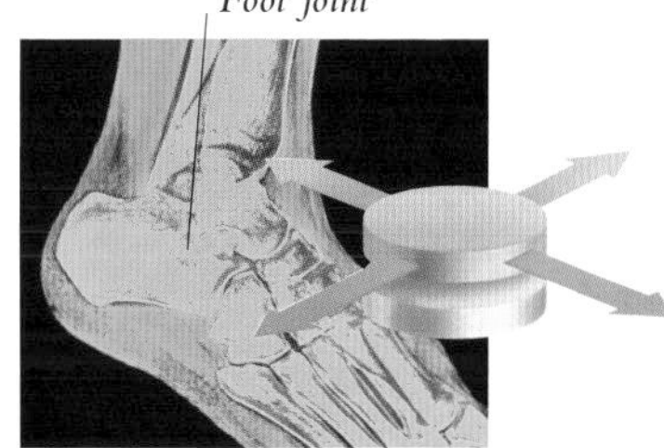

PLANE JOINT
Surfaces that are almost flat slide over each other, back and forth, and sideways. Some joints in the foot and wrist are plane joints.

SEMIMOVABLE AND FIXED JOINTS

Not all joints in the body are as freely movable as synovial joints. In semimovable joints, the articular surfaces are fused to a tough pad of cartilage that allows for only a little movement. The joints in the spine are semimovable, as is the joint at the base of the pelvis. Other joints allow for no movement at all and are fixed in place. The separate plates of bone in the skull allow for growth during childhood, but in adulthood, once growth is complete, they fuse together. The sacrum, in the lower spine, is another example of joints that are fused together. Here, individual vertebrae form a solid triangular unit, providing stability and support.

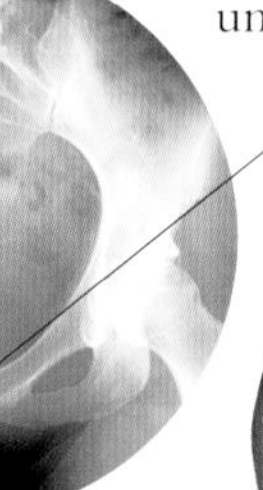

SEMIMOVABLE JOINT
The pubic symphysis joint links the front halves of the pelvis. It is semimovable, allowing for only a limited scope for movement at the front of the pelvis.

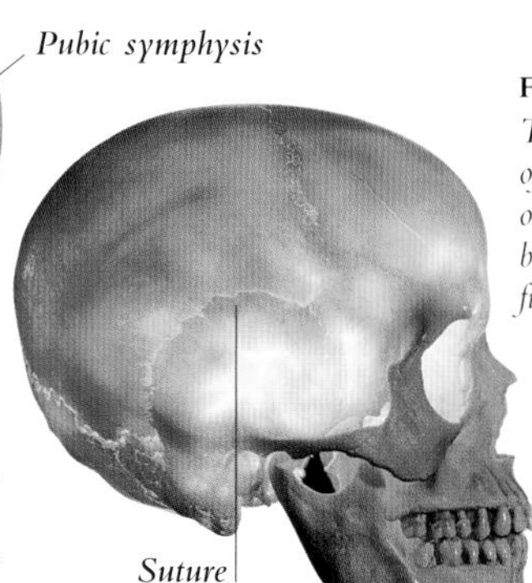

FIXED JOINTS
The suture joints of the skull hold or connect the bones of the skull firmly together.

KEEPING JOINTS STABLE

The body's synovial (freely movable) joints need to be kept stable while allowing maximum flexibility. Most joints are stabilized by ligaments, which are tough bands of fibrous, elastic connective tissue. External ligaments attach to the bones on either side of the joint, forming a fibrous capsule that completely encases the joint. The fibrous capsule protects the joint from damage or injury, while keeping it stable, yet flexible. As well as external ligaments, the knee joint has internal ligaments. These so-called cruciate ligaments are thick, fibrous bands that cross over inside the joint, linking the ends of the bones that form the joint. Some of the joints in the body, such as the ankle and wrist, also have thickened sheets of connective tissue called retinacula that wrap, like cuffs, around the tendons. The role of the retinacula is to hold the tendons in and stop them from bowing when the muscles contract and shorten.

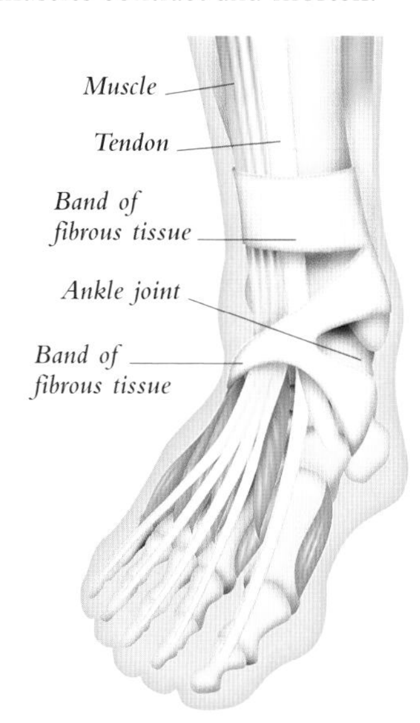

STABILIZING BANDS
Bands of fibrous connective tissue, called retinacula, wrap around the ankle to provide the joint with maximum stability.

Fact

Flexible joints

The vertebrae (bones in the back) are amazingly versatile, keeping us stable and upright when needed, and also allowing us to bend and arch backwards (see below) and forwards. Individually, the vertebrae are rigid and immovable, but as an interlocking group, they join together and work as a team to give the spine its flexibility. A network of ligaments connects the bones and the surrounding muscle, providing support for the joints and controlling movement.

Muscle

Patellar tendon
Crosses over the patella, which is embedded in it

Femur
Thighbone, the bone of the upper leg

Patella
Protective disc of cartilage, also called the kneecap

Synovial membrane
Produces synovial fluid

Pad of fat

Nerve

Artery

Vein

Attachment of patellar tendon to bone

Tibia
Shinbone, the larger of the two leg bones

KNEE JOINT
A typical synovial joint, the knee joint is held in place and stabilized by ligaments, both externally (see above) and internally (not shown). The joint is also protected by the patella, or kneecap, a disc of cartilage that fits on top of the joint.

INSIDE A JOINT

Synovial joints are structured so that they are movable, yet stable. The articular surfaces, where bone ends meet, are covered with smooth cartilage so that they can slide easily over each other. To ensure mobility, the ends of the bones are bathed in synovial fluid. This fluid also has a protective function and contains fats and proteins that nourish the cartilage covering the ends of the bones. The synovial membrane secretes and holds the fluid, reabsorbing and replenishing it continuously. The synovial membrane gradually thickens to form the outer fibrous capsule encasing the joint.

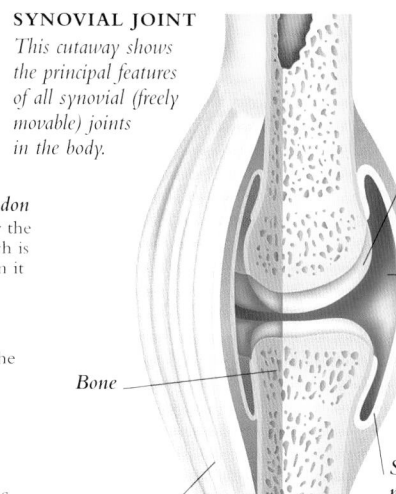

SYNOVIAL JOINT
This cutaway shows the principal features of all synovial (freely movable) joints in the body.

Articular cartilage
Covers bone ends, providing a smooth surface for bones to move over each other without causing damage

Synovial fluid
Lubricating fluid that bathes the bone ends, minimizing friction

Bone

Fibrous capsule
Forms a protective casing for the joint

Synovial membrane
Secretes lubricating synovial fluid

SHOCK ABSORBERS

In the knee, the wrist, and the spine, shock-absorbing fibrous discs positioned between the bones of the joints protect the bones from damage. These discs are called meniscuses in the knee and the wrist, and intervertebral discs in the spine. The structure of meniscuses can be described as being like a jam doughnut. The inner core consists of a jelly-like material comprising mainly water and collagen (a material composed of protein fibres); the outer part is made of tough, fibrous cartilage.

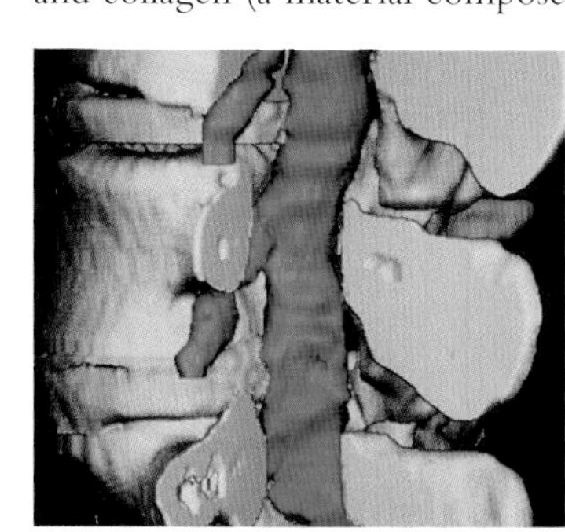

SPINAL DISCS
In this CT scan, parts of the vertebrae have been cut away to reveal the intervertebral discs (yellow), which absorb shock and prevent damage to the spine.

Health

Joint replacement

Many joints in the body that have been damaged can be replaced with a prosthesis. Damage to joints is often due to conditions that damage the cartilage around the bone, such as osteoarthritis. Joints that can be replaced include the hip, knee, and shoulder joints. Here, a prosthetic end has been inserted into the humerus bone of the upper arm, creating a new shoulder joint. Such replacements can last for many years.

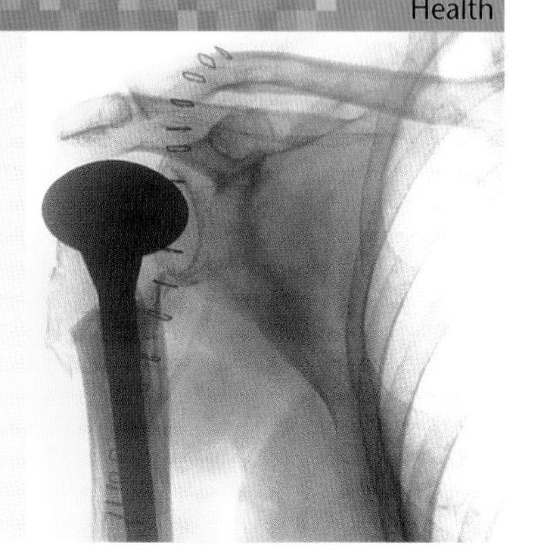

MUSCLES

Muscles make up the bulk of the body – accounting for half of its weight. They consist of tissue that contracts powerfully to move the body, maintain its posture, and work the various internal organs, including the heart and blood vessels. These functions are performed by three types of muscle – skeletal muscle, cardiac (heart) muscle, and smooth muscle (see below). Most of the muscle in the body is skeletal muscle. Usually, each end of a skeletal muscle is attached to a bone by a tendon, a flexible cord of fibrous tissue.

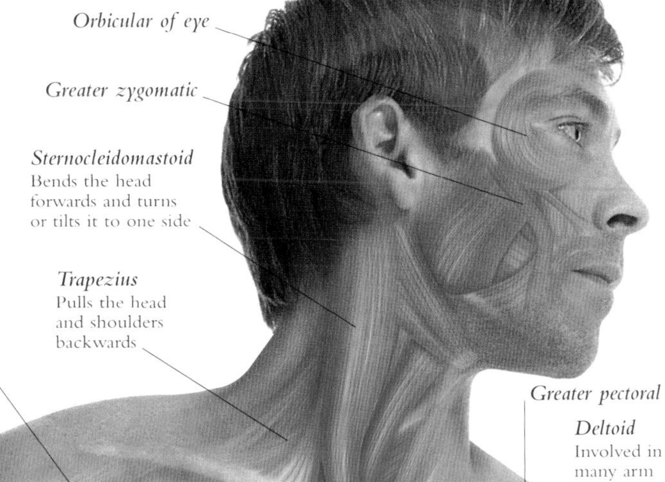

TYPES OF MUSCLE

There are three types of muscle in the body: skeletal, cardiac, and smooth. They each perform different roles and have differing structures. Skeletal muscle, which covers and moves the skeleton, consists of long cells, or fibres, that are able to contract quickly and powerfully. Cardiac (heart) muscle, is made up of short interlinked fibres capable of sustained rhythmic movement. Smooth muscle performs the unconscious actions of the body, such as propelling food along the digestive tract. Its fibres contract relatively slowly but they can continue in a state of contraction for long periods.

SKELETAL MUSCLE
The strong parallel fibres that form this type of muscle can contract quickly and powerfully, but only for a short time.

CARDIAC MUSCLE
Short, branching, interlinked fibres form a network within the wall of the heart. Healthy cardiac muscle contracts rhythmically and continuously in order to pump blood around the whole of the body.

SMOOTH MUSCLE
These short, spindle-shaped fibres are thinner than skeletal muscle fibres, and form sheets of muscle. Smooth muscle can contract for long periods.

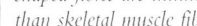

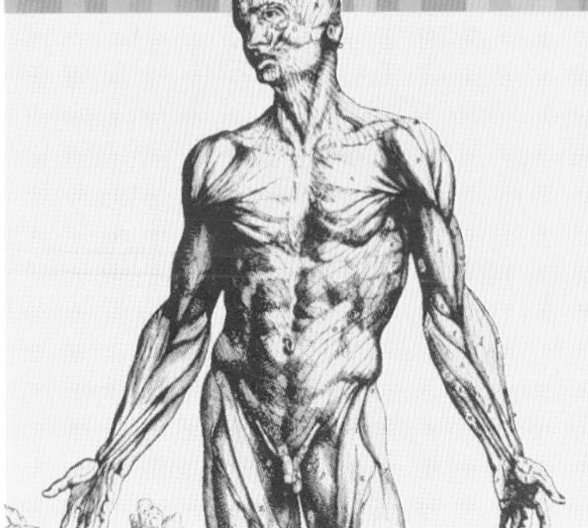

Discovering muscles

Belgian scholar Andreas Vesalius (1514–1564) uncovered the intricate layers of the body's skeletal muscles by careful and accurate dissection. He established the science of anatomy, based on observation and realism. The famous engraving shown here was first published in 1543 in Vesalius' book entitled *On the Structure of the Human Body*, in which he describes cutting away "the skin together with the fat, and all the sinews, veins and arteries existing on the surface".

SKELETAL MUSCLE

The body's skeletal muscles move the body and joints by contracting. In addition, they maintain a steady tension, or tone, that gives the body the support it needs to sustain posture. Our vast range of facial expressions, which is a significant means of communication, is also facilitated by our skeletal muscles. There are over 600 skeletal muscles in the body, in a variety of shapes and sizes – from large triangular slabs of muscle, such as the deltoid in the shoulder, to long, thin strips, such as the sartorius, which curves around from the hip to the inside knee. Skeletal muscle can contract quickly and powerfully. The shape, size, and length of a muscle all have a bearing on the strength with which it contracts, and this influences the amount of force it can generate. Muscle contraction is stimulated by nerve impulses, sent along pathways linking the brain to the muscle tissue.

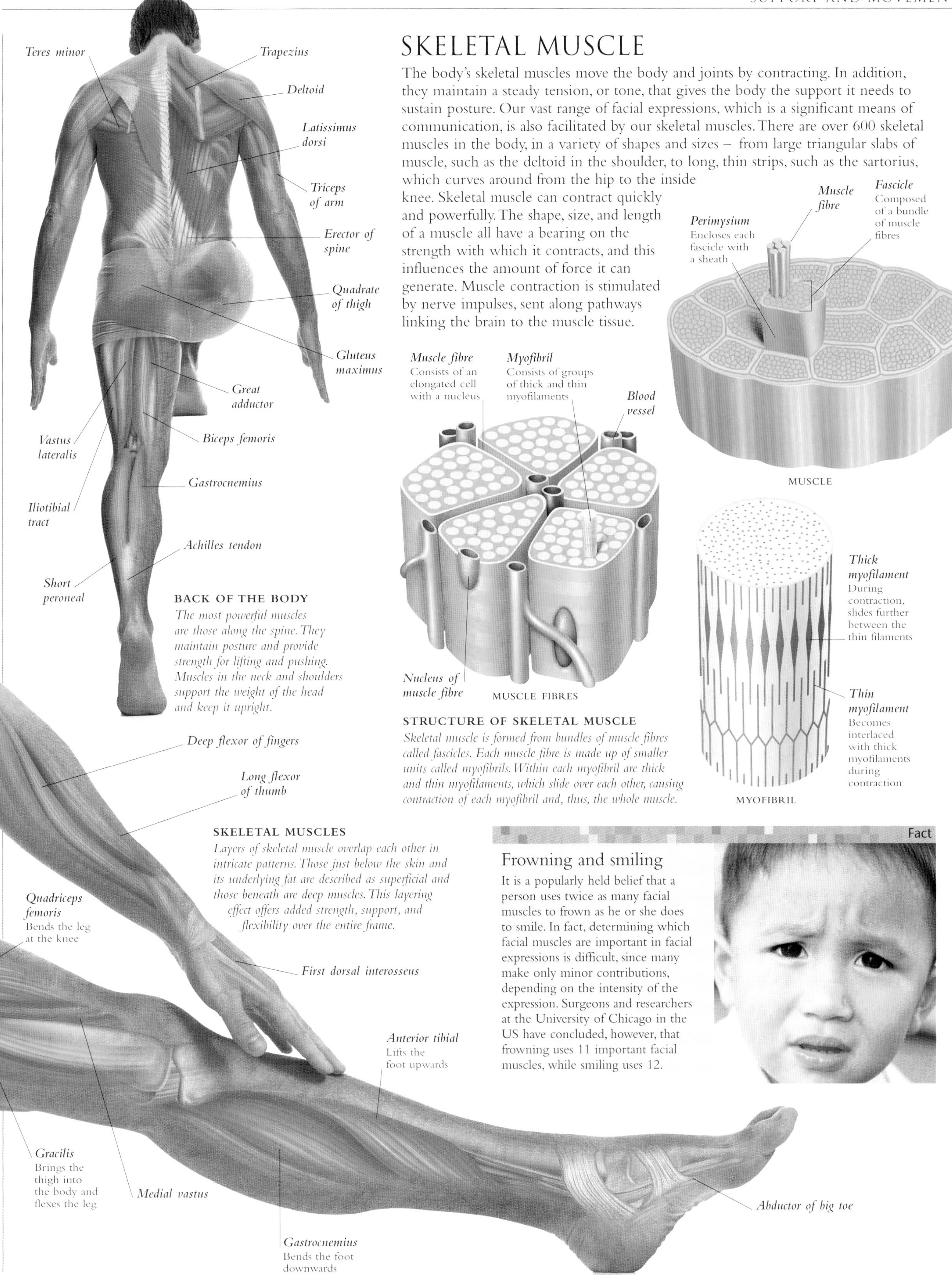

BACK OF THE BODY

The most powerful muscles are those along the spine. They maintain posture and provide strength for lifting and pushing. Muscles in the neck and shoulders support the weight of the head and keep it upright.

STRUCTURE OF SKELETAL MUSCLE

Skeletal muscle is formed from bundles of muscle fibres called fascicles. Each muscle fibre is made up of smaller units called myofibrils. Within each myofibril are thick and thin myofilaments, which slide over each other, causing contraction of each myofibril and, thus, the whole muscle.

SKELETAL MUSCLES

Layers of skeletal muscle overlap each other in intricate patterns. Those just below the skin and its underlying fat are described as superficial and those beneath are deep muscles. This layering effect offers added strength, support, and flexibility over the entire frame.

Fact

Frowning and smiling

It is a popularly held belief that a person uses twice as many facial muscles to frown as he or she does to smile. In fact, determining which facial muscles are important in facial expressions is difficult, since many make only minor contributions, depending on the intensity of the expression. Surgeons and researchers at the University of Chicago in the US have concluded, however, that frowning uses 11 important facial muscles, while smiling uses 12.

TENDONS

Muscles are linked to bones by tendons, which are cord-like structures made of collagen, a tough, fibrous protein. Tendons, unlike muscles, do not stretch but they have some flexibility. Where they join to a bone, tendon fibres pass through the bone's outer membrane, the periosteum, and embed themselves in the bone tissue. The link is extremely strong, and tendons resist great tension without snapping. Some tendons, including those in the hands and feet, are encased in fibrous capsules called synovial sheaths. The sheaths secrete a lubricating fluid that protects the tendons from friction where they move against the bone.

TENDON FIBRES
Tendons are largely made up of collagen, a fibrous connective tissue. The fibres of a tendon are arranged in parallel bundles (as shown in the photograph), forming a cord-like structure of great strength.

HOW MUSCLES WORK

Skeletal muscles typically connect two bones, stretching from one to the other across a joint. An individual muscle produces movement when it contracts (shortens) and pulls on the bone to which it is attached. Muscles can only pull bones towards or away from each other; they cannot push. For this reason, many muscles are arranged in pairs, on either side of a joint, working in opposition to move body parts. While one muscle or muscle group contracts, pulling a bone in one direction, the muscle or muscle group on the opposite side of the joint relaxes. In order to reverse the movement, the first muscle relaxes while its counterpart contracts. Examples of paired muscles are the biceps and triceps in the upper arm. When the body is not moving, all muscles are held in a state of partial contraction. This natural tension, muscle tone, maintains posture.

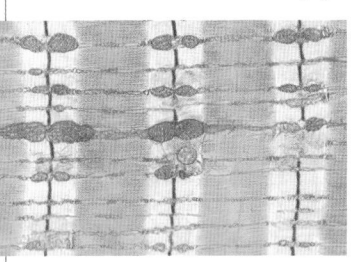

MUSCLE FILAMENTS
The filaments that form skeletal muscle are arranged in a repeating pattern of stripes. In order to move the body, the muscle fibres need to contract and relax.

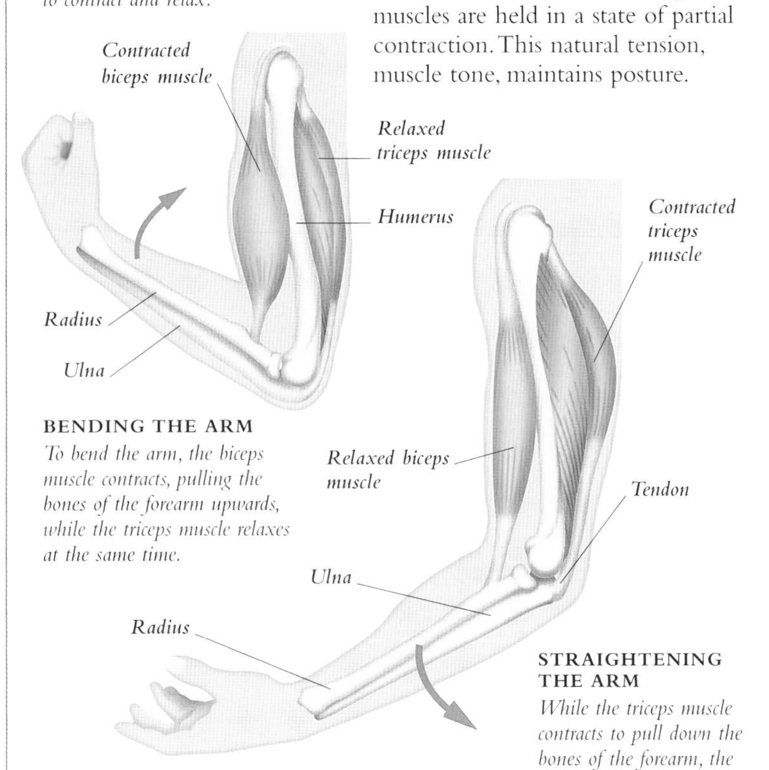

BENDING THE ARM
To bend the arm, the biceps muscle contracts, pulling the bones of the forearm upwards, while the triceps muscle relaxes at the same time.

STRAIGHTENING THE ARM
While the triceps muscle contracts to pull down the bones of the forearm, the biceps relaxes.

LEVERS

Muscles pull on bones to make the body move according to the same principles that operate mechanical lever systems. A lever is a rigid bar that has one pivot point, the fulcrum; force applied to one part of the lever is transferred through the fulcrum to a weight (resistance point) on another part of the lever. Translated into bodily terms, this means that the bones serve as levers; the muscles apply force; the part of the body to be moved, such as a limb, provides resistance; and the joints at which bones meet function as fulcrums. There are three types of lever system in the body: first, second, and third class. These are classified according to the relative positions of the fulcrum, the force applied, and the weight being resisted.

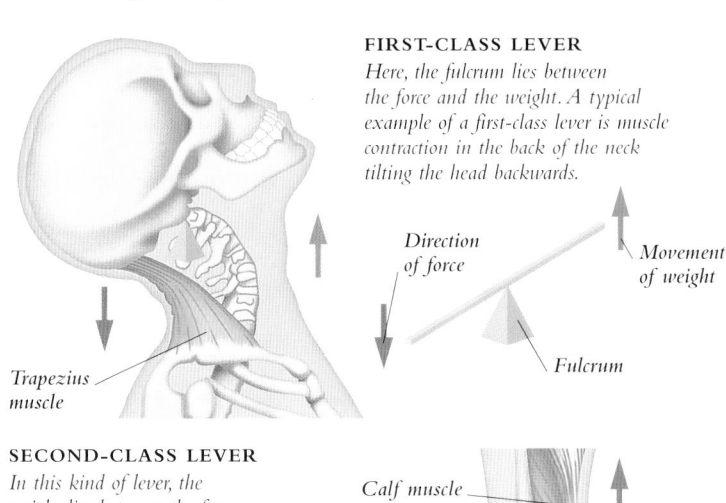

FIRST-CLASS LEVER
Here, the fulcrum lies between the force and the weight. A typical example of a first-class lever is muscle contraction in the back of the neck tilting the head backwards.

Direction of force
Movement of weight
Fulcrum
Trapezius muscle

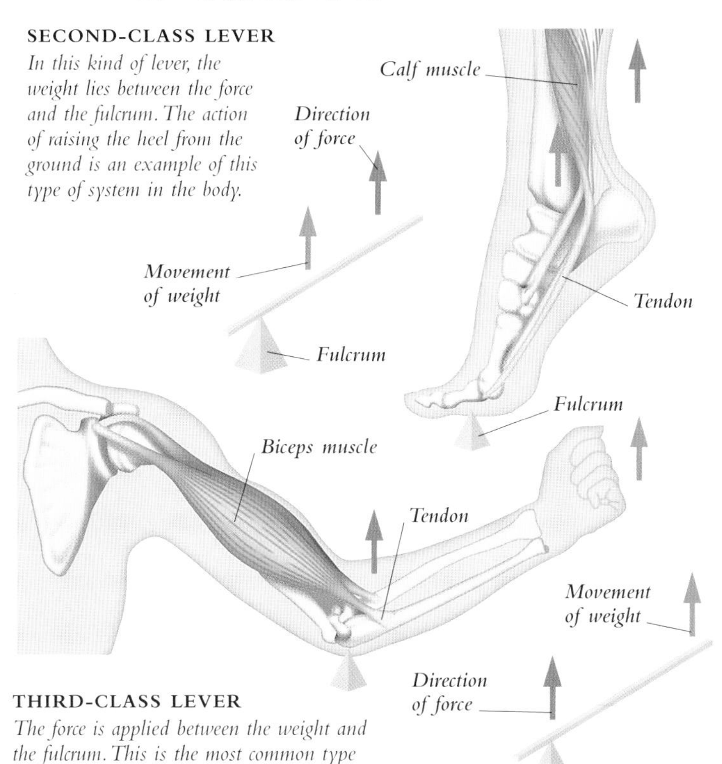

SECOND-CLASS LEVER
In this kind of lever, the weight lies between the force and the fulcrum. The action of raising the heel from the ground is an example of this type of system in the body.

THIRD-CLASS LEVER
The force is applied between the weight and the fulcrum. This is the most common type of lever in the body, an example of which is bending the elbow by contracting the biceps.

Fact

Bodybuilding

The size and strength of the individual fibres in skeletal muscles can be gradually increased by physical exercise. Regularly performing multi-repetition exercises, especially using weights, is an effective way of developing well-defined muscles. A dangerous and illegal short-cut to bodybuilding is the abuse of certain drugs, anabolic steroids, that promote the growth of muscle tissue. Long-term use of these drugs can produce side-effects, including liver damage and reduced fertility. Anabolic steroids are prohibited by sporting and athletic authorities worldwide.

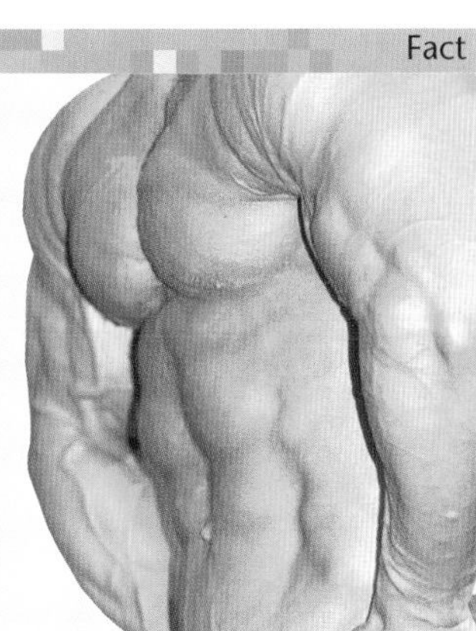

Health

Muscle cramps

During vigorous physical exercise, sudden muscle cramps may develop. A common reason for these painful spasms is the accumulation of a waste product, lactic acid, in muscle cells. This waste builds up when the body is short of oxygen. Carried to muscle fibres in fine, branching capillaries (*see below*), oxygen is essential to muscle activity. If muscles run out of oxygen, for instance in a race, they make energy without it, which produces lactic acid.

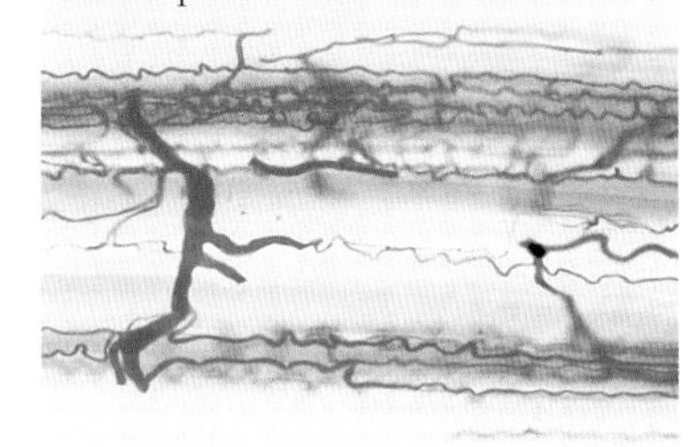

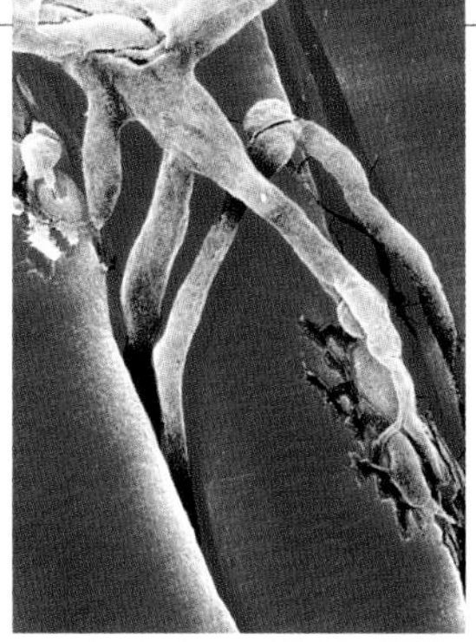

NERVE JUNCTION
Impulses from the brain, triggering muscle action, reach muscle fibres through the endings of nerves (green).

NERVOUS SYSTEM CONTROL

Movement of the body depends not just on mechanical interaction of muscles and bones, but on signals from the brain and nerves. Contraction of skeletal muscles is often involuntary, but it is also likely to be the result of conscious thought. Once our brain has made the decision to move something – for example, to take a step forwards or bend the arm – it sends out electrical signals to muscles along nerve pathways. When the signals arrive at the appropriate muscle, the filaments within the muscle fibres respond by contracting. If signals from the brain cease, the muscle fibres are no longer stimulated and the muscle relaxes. Another link between the brain and body movement is an internal monitoring system, proprioception. Proprioceptors are types of sensory receptors, located in muscles and tendons, that collect information about the degree of stretch of muscles and tendons around the body. This information, which is passed to the central nervous system (the brain and spinal cord) helps to give us our sense of balance and our awareness of the position of various parts of the body in relation to each other.

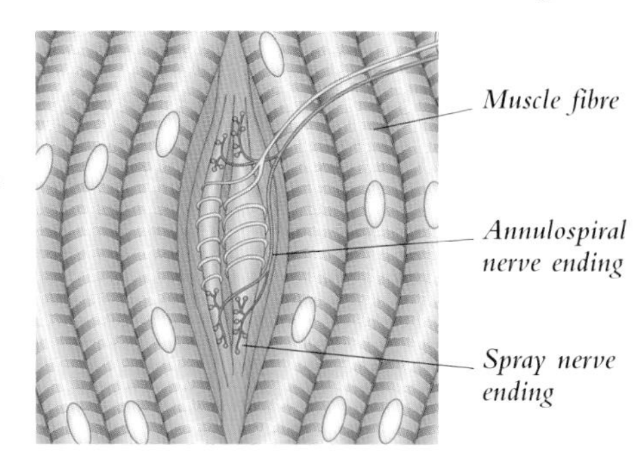

PROPRIOCEPTORS
Two types of muscle proprioceptor are annulospiral sensory nerve endings, which wind around the fibres, and spray nerve endings, which lie on top of the fibres.

FAST AND SLOW MUSCLE FIBRES

Skeletal muscles are made up of two types of fibre: fast and slow. Fast muscle fibres contract rapidly, enabling someone to produce a burst of intense activity, such when making a sudden sprint or lifting a heavy weight. However, these fibres also tire quickly. Slow muscle fibres contract more slowly but can keep going for a long time. For example, a marathon runner uses slow fibres to sustain a steady, untiring pace over a long distance. The fibre types differ in the way they produce energy for muscle contraction. To be able to function, slow fibres need oxygen, obtained from circulating blood. The cells that make up slow fibres contain many mitochondria; these are structures that use nutrients and oxygen taken in by the body to create fuel for activity. Fast fibres have fewer and smaller mitochondria than slow fibres, and are able to produce energy without oxygen, although only in relatively small amounts. In most people, the proportion of fast and slow fibres in the skeletal muscles is about equal. However, some top athletes do seem to have a greater percentage of one type of fibre over the other, a genetic predisposition which possibly contributes to their particular talents. A sprinter or a basketball player may have more fast fibres, while a long-distance runner has more slow fibres.

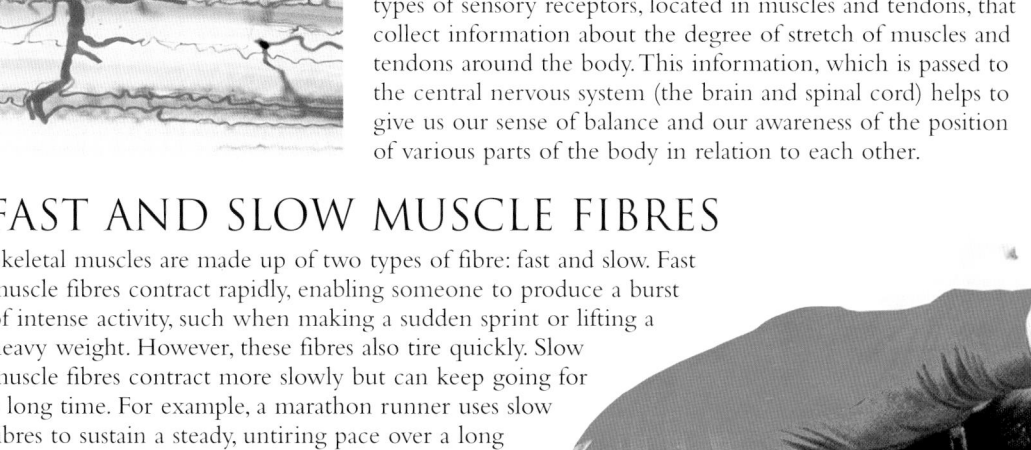

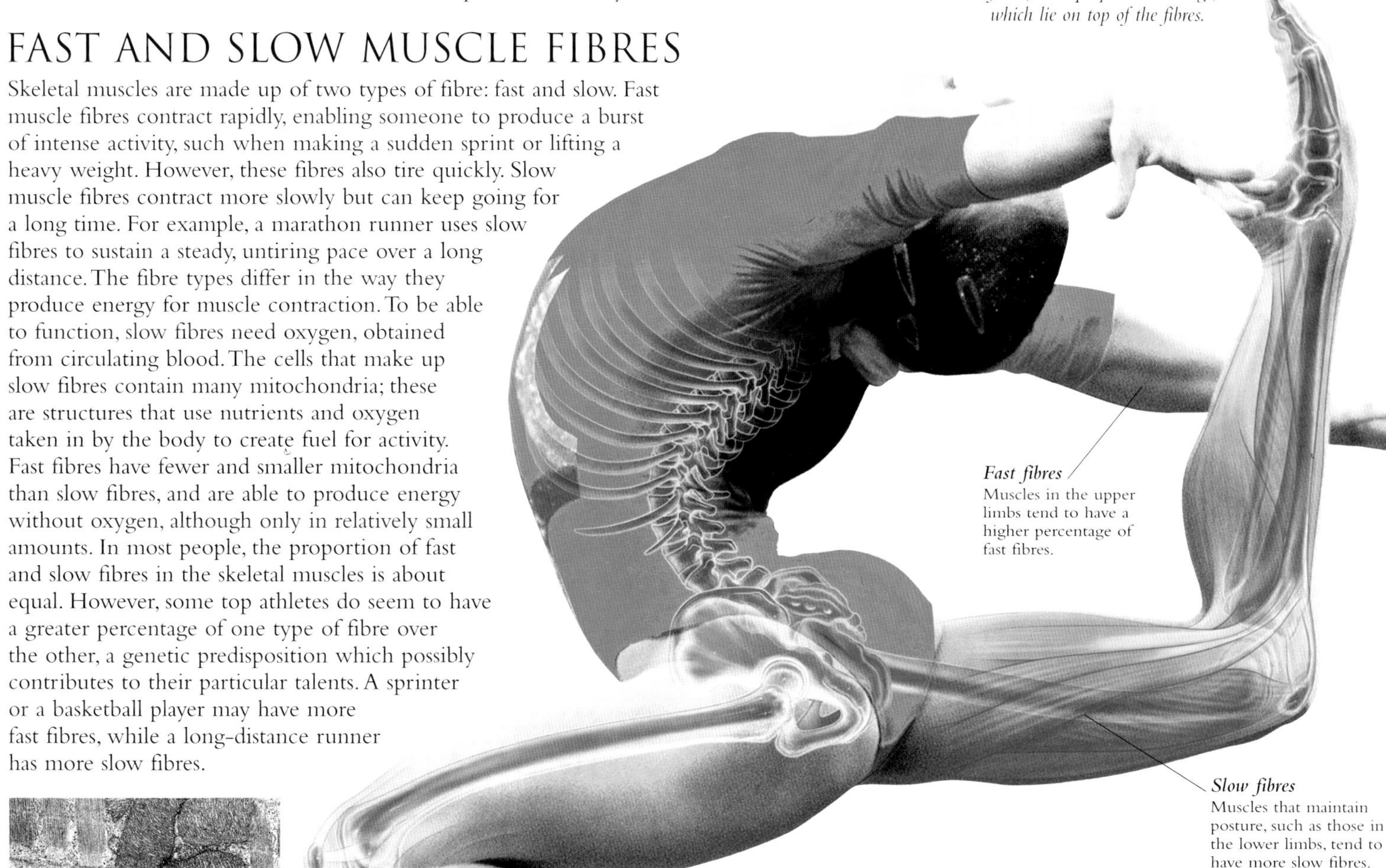

Fast fibres
Muscles in the upper limbs tend to have a higher percentage of fast fibres.
Slow fibres
Muscles that maintain posture, such as those in the lower limbs, tend to have more slow fibres.

TYPES OF FIBRE
Like everyone, this gymnast has a mix of fast and slow fibres in her muscles. She needs the fast type to perform rapid movements and the slow fibres for more sustained exercises.

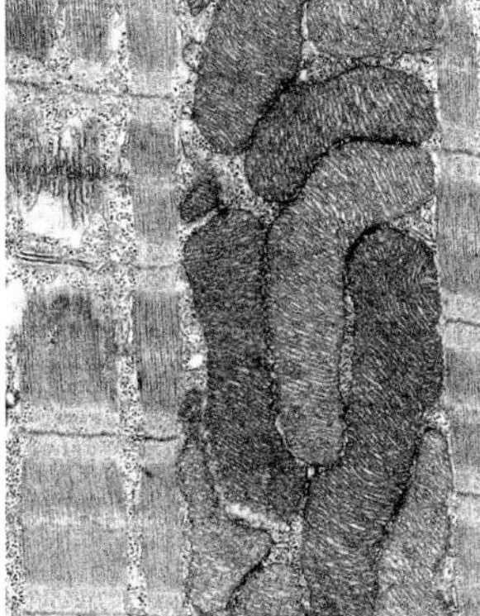

ENERGY PRODUCERS
Large numbers of energy-producing mitochondria (centre of photograph) occur in slow muscle fibres.

Fact

Speed limits

All mammals, including humans, move by using the same mechanical principles. However, specialized adaptations between one species and another make a vast difference to physical performance. No amount of training could make a human win a race against a cheetah, the fastest animal on earth. This animal's long legs and spine, and the rapidity with which its muscles contract help it to reach speeds of 100km/h (62mph).

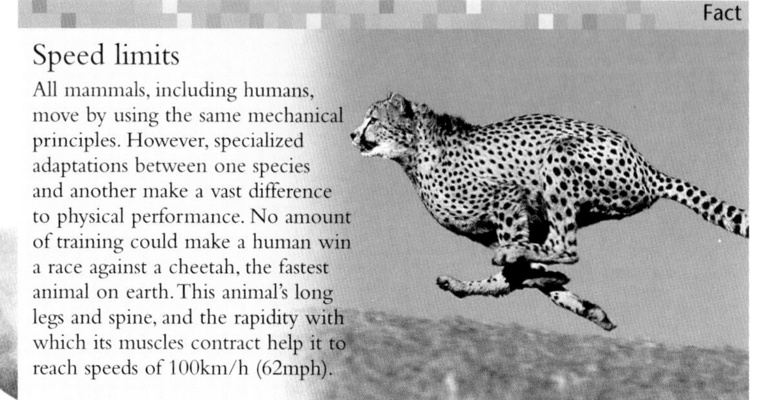

USING OXYGEN
A swimmer breaks the surface of the water for a gulp of air. Humans cannot survive for more than a few minutes without replenishing their oxygen supply.

BREATHING AND CIRCULATION

Animals need a constant supply of oxygen; without this vital gas, they would die quickly. Oxygen keeps body cells working by releasing energy from the nutrients we obtain from food. Breathing draws in plenty of oxygen, but an efficient transport system is needed to carry the oxygen around the body. This role is performed by the heart and the circulating blood. Another important function of breathing and blood circulation is to remove the waste gas, carbon dioxide, produced by body cells.

The cycling of oxygen and carbon dioxide between the body and the atmosphere – respiration – is not just a matter of breathing in and out but involves every cell in the body. This complex process, which for the most part takes place without any conscious effort, needs the coordination of many body parts and functions. The brain, airways and lungs, heart, blood, and blood vessels are all essential components.

AIR INTAKE

No one can survive for more than a few minutes without taking in air. Most of the time we are not aware of the action of breathing, although we draw breath over 20,000 times a day. The rhythmic, reflex action, as the lungs expand and shrink, is controlled by the brain. Breathing automatically adjusts in depth and rate in order to meet different levels of exertion. Breathing becomes noticeable only if it is difficult or we try to take deliberate control.

Every breath of air we inhale is filtered, warmed, and moistened as it flows through the cavities of the nose. After travelling down the pharynx (throat) and trachea (windpipe), inhaled air passes along ever-branching pathways until it is deep within the spongy tissues of the lungs. The moist interior of the lungs provides a damp surface through which oxygen can diffuse easily to reach the bloodstream. There are clusters of tiny air sacs, alveoli, in the lungs that together provide a huge surface area. This arrangement maximizes oxygen intake. Waste carbon dioxide leaves the body by the reverse route, entering the lungs through the alveoli. This gas is removed through the airways as we exhale.

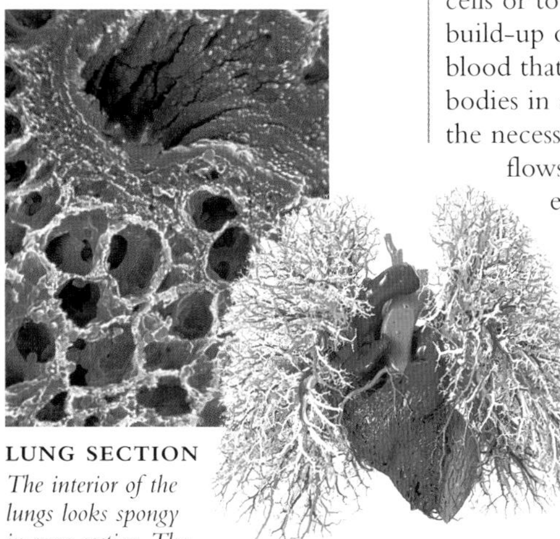

LUNG SECTION
The interior of the lungs looks spongy in cross-section. The tiny airways seen here are bronchioles (top of photograph) and alveoli (bottom).

BRONCHIAL TREE
This resin cast shows how the air passages, bronchi, in the lungs branch to form an intricate network.

VITAL EXCHANGE

Human bodies are so large that simple gas diffusion through the lungs is not fast enough either to get oxygen to all our billions of cells or to prevent a dangerous build-up of carbon dioxide. The blood that travels around our bodies in a double circuit provides the necessary service. As blood flows through the lungs an exchange of gases takes place through the thin walls of the alveoli. The cells in blood collect oxygen and deposit carbon dioxide. From the lungs, oxygenated blood goes to the heart, which pumps it around the entire body. The oxygenated blood is carried through arteries, the largest of our blood vessels, then through smaller and smaller channels until it has reached every tissue. Once in the tissues, blood releases its supply of oxygen to cells that need it and picks up cell waste, carbon dioxide. To complete the circuit, the blood returns to the heart through a network of veins. The cycle then repeats as blood is pumped back to the lungs to rid itself of carbon dioxide and pick up more oxygen.

BLOOD CIRCULATION

Under a microscope, it can be seen that blood is made from various types of cells floating in a watery fluid and that its colour comes from millions of round, red discs. These discs are red blood cells; they have special properties that enable them to collect oxygen in the lungs and release it in body tissues. The other components of blood take no part in oxygen transport. The fluid in which the cells float (plasma) carries nutrients; white blood cells help to protect the body from disease; and platelets are involved in blood clotting.

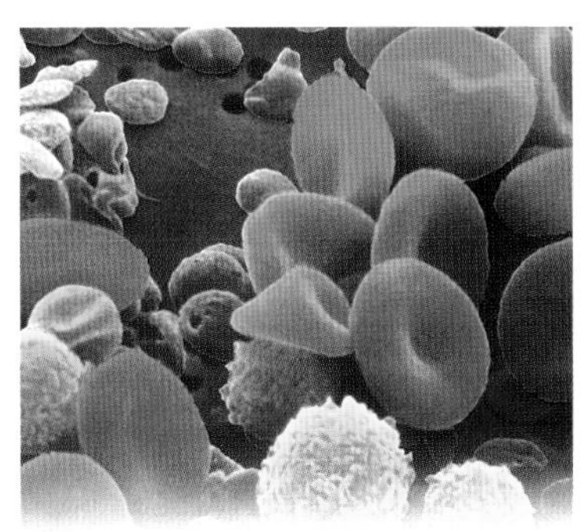

BLOOD CELLS
A sample of blood shows a variety of cells: red cells carry oxygen, white cells fight disease, and tiny platelets aid clotting.

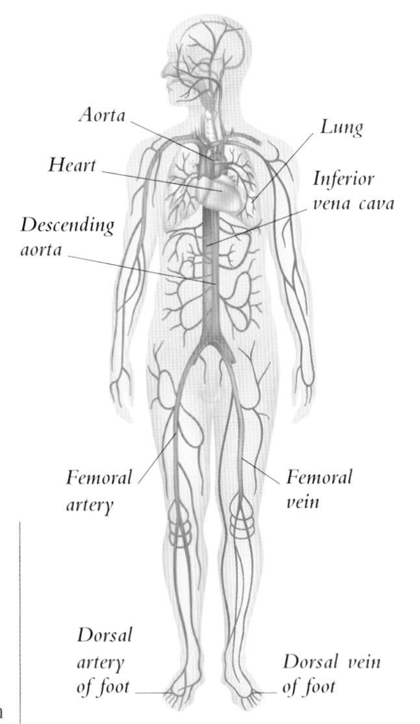

OXYGEN CYCLE
Blood containing oxygen collected in the lungs is pumped into the arteries (red) by the heart and around the body. Oxygen-poor blood drains back to the heart through the veins (blue) and is pumped to the lungs.

To get enough oxygen to all of the body's organs, all of our blood must pass through the lungs and around the body at least once a minute. Keeping this nonstop circulation going means that blood has to travel fast and therefore must be actively forced through the blood vessels. The heart is the pump that keeps blood flowing. Situated between the lungs, tilted a little towards the left side of the body, this hollow, muscular sac works tirelessly. Expanding and contracting with a steady rhythm, the heart fills with blood and then pumps it out again to the lungs and the body's labyrinth of blood vessels. The special heart muscle beats of its own accord, producing the characteristic sound heard in a stethoscope. Although heart action is automatic, the rate at which the heart beats is regulated by signals from the brain.

Fact

Powerful lungs

Exercise increases the volume of the alveoli (air sacs) in the lungs, which means that more oxygen gets into the body with every breath. Because of their rigorous training, professional cyclists, such as Lance Armstrong (right), are known by sports physicians for their increased lung capacity. Some professional sports people are recorded as taking in up to 8 litres (14pt) of air a minute – people of ordinary fitness levels take in 5–6 litres (9–10pt).

TAKING A BREATH

Breathing has two purposes: it gets oxygen into the body and removes carbon dioxide waste produced by body cells. Every minute, about 5–6 litres (9–10pt) of air pass into and out of the lungs. Oxygen from air entering the lungs diffuses into the bloodstream and is pumped through the circulation to all of the body's cells. In the body's tissues, the oxygen is exchanged for carbon dioxide waste, which is then returned, through the circulation, to the lungs to be exhaled.

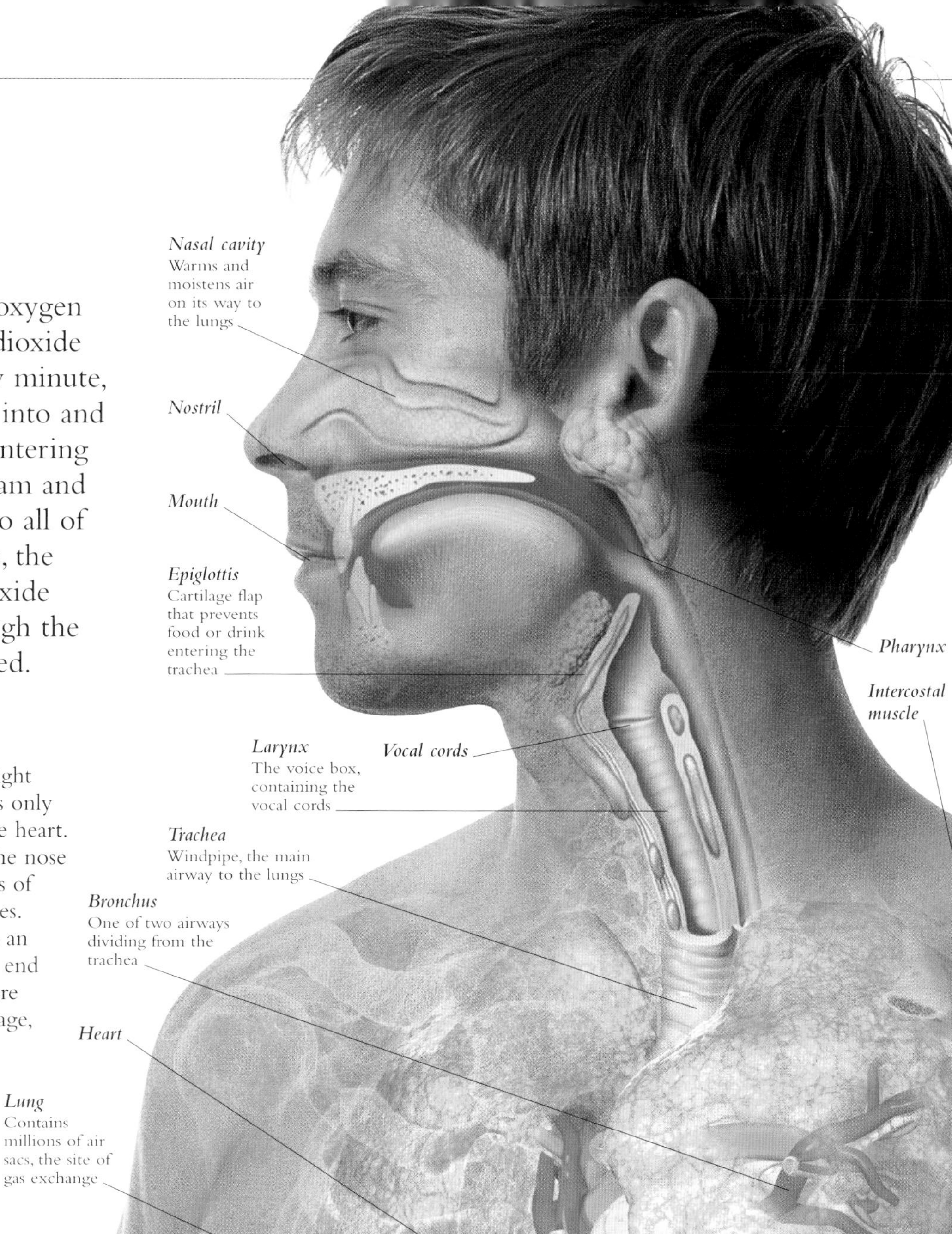

THE RESPIRATORY TRACT
The nasal cavity and the pharynx (throat) form the upper respiratory tract. The larynx (voice box), trachea (windpipe), bronchi (air passages), and lungs make up the lower part.

THE LUNGS

Healthy lungs are pink, soft, and spongy. The right lung is divided into three lobes, but the left has only two lobes because it shares chest space with the heart. Air moves into and out of the lungs through the nose and mouth, the trachea (windpipe), and a series of other air passages called bronchi and bronchioles. These airways branch out inside the lungs into an extensive, tree-like network of small tubes that end in alveoli (tiny air sacs). The lungs themselves are easily damaged, but three structures – the ribcage, spine, and a double-layered membrane, called the pleura – protect them from harm.

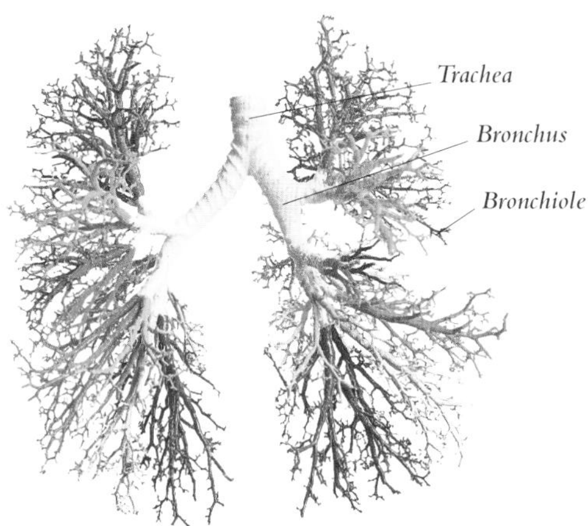

AIR PASSAGES IN THE LUNGS
The largest airway in the respiratory system is the trachea. This branches into the two smaller bronchi which in turn divide to become bronchioles.

Health

Coughs and sneezes

The noisy, explosive force of coughing or sneezing is an effective way of clearing the airways and preventing harmful and irritating substances, such as dust, pollen, or mucus, entering the lungs and causing damage. Sneezing and coughing are often symptoms of viral infections such as the common cold and flu. Coughing can also indicate more serious disorders, such as pneumonia or damage to airways and lung tissues caused by smoking.

BODY

GAS EXCHANGE

The tiny air passages in the lungs end in millions of alveoli – thin-walled air sacs – where gases are exchanged between the air and the blood. As oxygen enters an alveolus it dissolves in the moist lining and diffuses across the thin wall into a neighbouring capillary (tiny blood vessel). This happens rapidly: at a restful breathing rate, the blood in the capillaries comes into contact with air in an alveolus for only about 0.75 seconds but is fully oxygenated after about a third of this time. Once oxygen is in the blood, it binds with haemoglobin in red blood cells (a small amount of oxygen also travels freely in the blood) and is transported to the heart, where it is pumped to the body tissues. Carbon dioxide follows the opposite path to oxygen, and travels about 20 times as fast; it diffuses out of the capillaries and into the alveoli, where it is exhaled from the lungs.

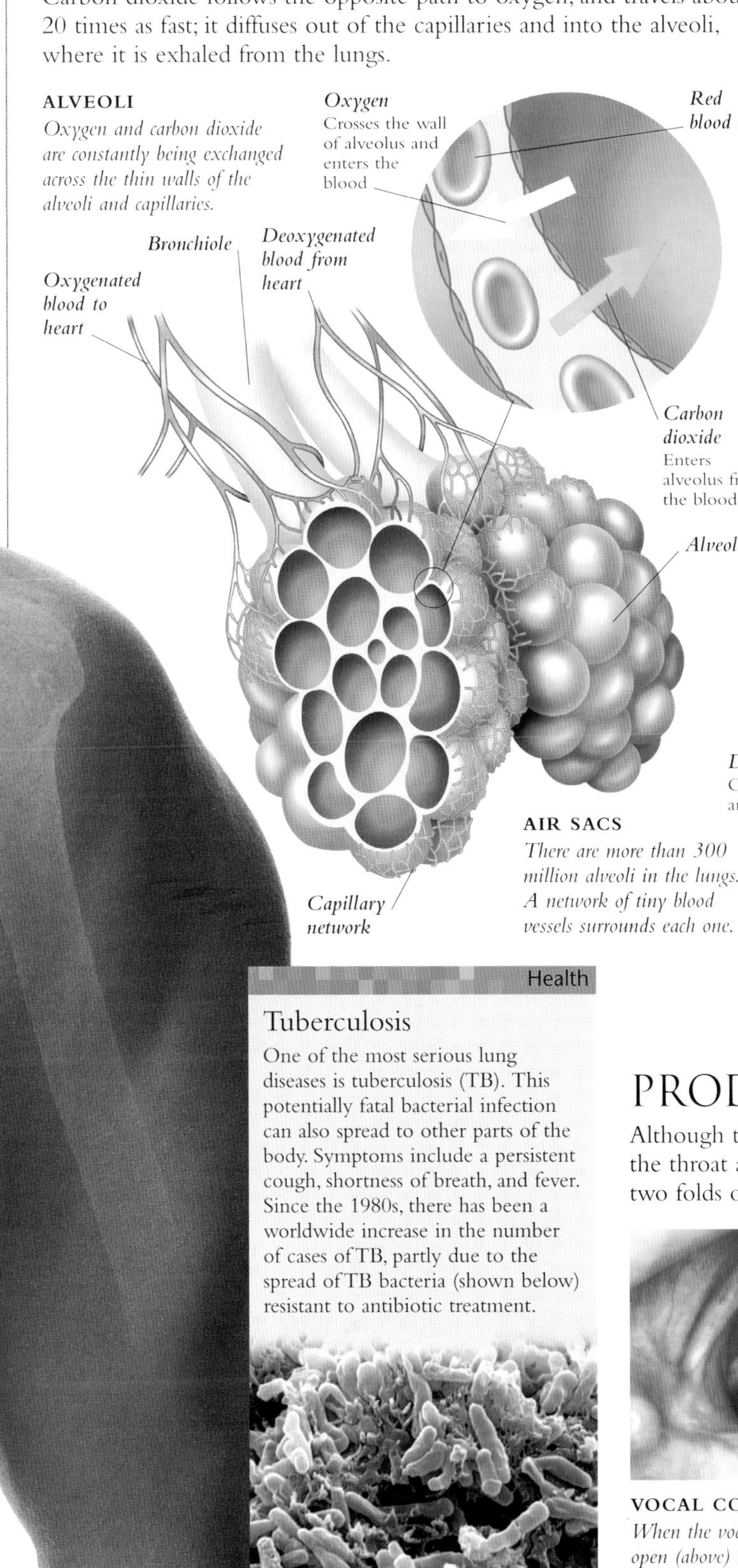

ALVEOLI
Oxygen and carbon dioxide are constantly being exchanged across the thin walls of the alveoli and capillaries.

AIR SACS
There are more than 300 million alveoli in the lungs. A network of tiny blood vessels surrounds each one.

Fact

Free diving

In the sport of free diving, people train themselves mentally and physically to survive underwater on a single lungful of breath. The record for static free diving (staying still underwater for as long as possible) is more than 6 minutes for women and more than 8 minutes for men. Pearl divers in the South Pacific sometimes swim vigorously underwater without breathing equipment for up to 2 minutes at depths of 40m (130ft).

GETTING AIR INTO THE LUNGS

When we breathe in, muscles work to expand the chest. The largest chest-expanding muscle is the diaphragm, a sheet of tissue that lies under the ribcage. At rest, the diaphragm arches upwards like a dome, separating the chest from the abdomen. When the diaphragm contracts, it flattens, pushing down the abdominal organs and increasing the space within the chest cavity. The intercostal muscles between the ribs also pull the chest up and out. The lungs increase in size, their internal pressure drops and air rushes in. Exhaling does not usually involve any work by the body. The diaphragm and intercostal muscles simply relax and return to their resting positions. This causes the lungs and chest wall to recoil like elastic. As they do so, the pressure in the lungs increases and air is forced back into the atmosphere.

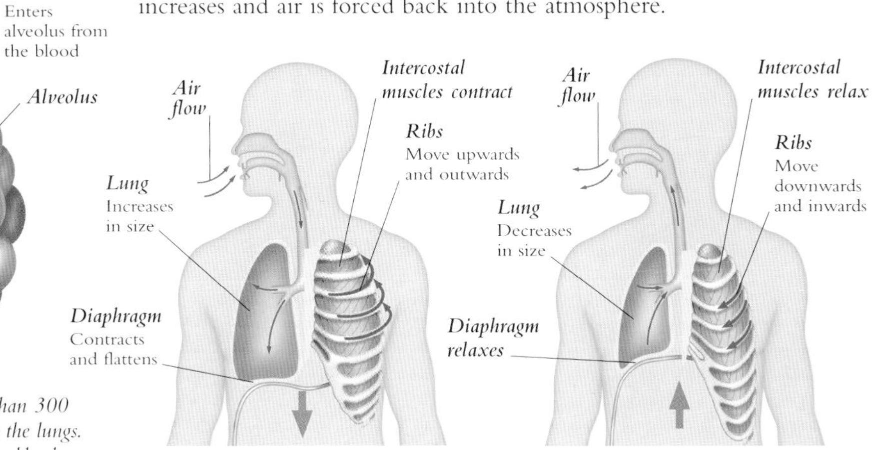

INHALING
The diaphragm contracts and flattens and the intercostal muscles also contract. As the ribcage expands, pressure in the lungs drops and air rushes in.

EXHALING
The diaphragm returns to its dome-like shape and the intercostal muscles relax. Pressure in the lungs rises and air is forced out.

Health

Tuberculosis

One of the most serious lung diseases is tuberculosis (TB). This potentially fatal bacterial infection can also spread to other parts of the body. Symptoms include a persistent cough, shortness of breath, and fever. Since the 1980s, there has been a worldwide increase in the number of cases of TB, partly due to the spread of TB bacteria (shown below) resistant to antibiotic treatment.

PRODUCING SOUNDS

Although the lungs evolved to deliver oxygen to the body, the passage of air through the throat also has the effect of producing sound. This effect is due to the vocal cords, two folds of mucous membrane within the larynx (voice box). When we speak, muscles in the larynx contract, pulling the vocal cords close together, and air is expelled from the lungs. The air forces its way through the small gap left between the vocal cords, causing the membranes to vibrate and creating the sounds of human speech. When the muscles relax, the vocal cords move apart and no sound is made. The length and tightness of the vocal cords affect the pitch of the voice – because men's vocal cords are longer and vibrate more slowly than women's, the male voice sounds deeper. The loudness of the voice depends on the force with which air pushes between the vocal cords.

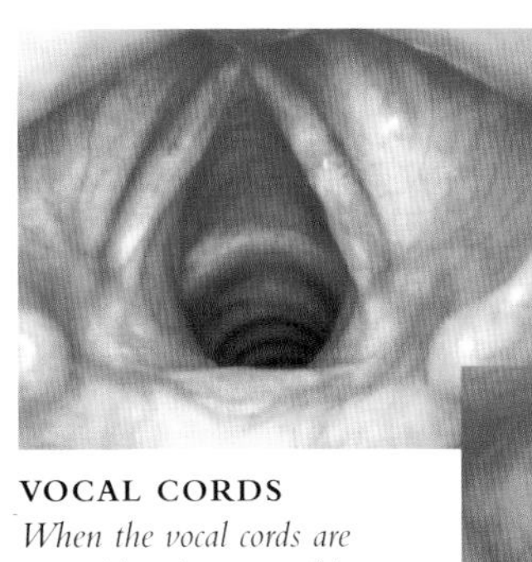

VOCAL CORDS
When the vocal cords are open (above) no sound is made, but when they are closed (right, air vibrates across them to produce the sound of the human voice.

CIRCULATION

Blood vessels carry nutrient- and oxygen-rich blood from the heart to the rest of the body and return used blood for replenishment. The three main types of blood vessel are arteries, veins, and capillaries. The largest artery, the aorta, emerges from the heart and branches into a network of progressively smaller arteries that take blood all over the body. The smallest arteries form capillaries, where exchange of oxygen and nutrients for carbon dioxide and cell waste products occurs. The capillaries then join a network of tiny veins that merge into larger veins as they return blood to the heart.

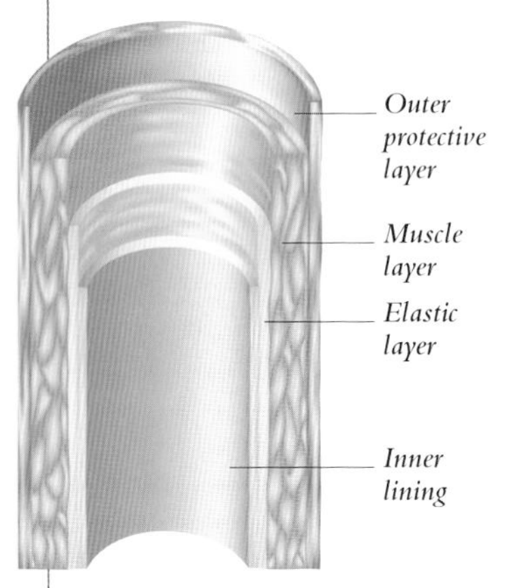

STRUCTURE OF ARTERY
Arteries have thick, muscular walls that can resist the waves of high-pressure blood from the heart.

ARTERIES

Arteries are the primary supplier of blood to the body – taking blood away from the heart, picking up oxygen from the lungs, and then transporting the oxygen-rich blood to different areas of the body. They are large vessels with thick, elastic muscular walls. Some of the main arteries, such as the carotid (in the neck), aid the pumping of blood from the heart by contracting rhythmically, pulsing blood through the body. The largest artery in the body, the aorta, which carries blood out of the left side of the heart, is particularly elastic and is almost as wide as a garden hose.

VEINS

Veins are responsible for taking deoxygenated blood from the rest of the body back to the heart. They do not have to deal with such high blood pressures as arteries because they are carrying blood towards the heart rather than away from it. Consequently, they have much thinner walls, with less elasticity and fewer muscle fibres. This structure means that veins are often flatter than muscular arteries, which allows surrounding muscles to act on them, helping to squeeze deoxygenated blood along. The main veins in the body (such as the jugular vein in the neck and the main veins in the legs) contain one-way valves that keep blood flowing towards the heart but prevent it from going back the other way.

Outer layer
Inner lining
Valve flap
Stops blood from flowing the wrong way
Muscle layer

STRUCTURE OF VEIN
Veins have thin walls that enable them to hold large volumes of blood. Large veins contain valves.

CAPILLARIES

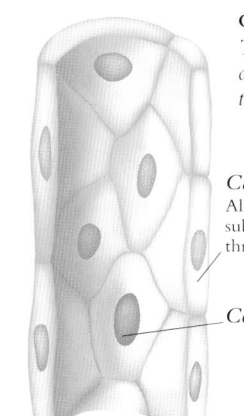

Oxygen-carrying arteries divide into smaller blood vessels called arterioles, which themselves divide into tiny vessels known as capillaries. The capillaries join up to form venules, which in turn join to form veins. The veins carry deoxygenated blood back to the heart. Capillaries have a very important role in circulation, because it is in their web-like beds that the exchange of oxygen and nutrients for waste occurs. The capillaries are so small and delicate that their walls are only one cell thick – it takes 10 capillaries to equal the thickness of a human hair. It is the pores and gaps in the capillary walls that allow nutrient and waste exchange to take place.

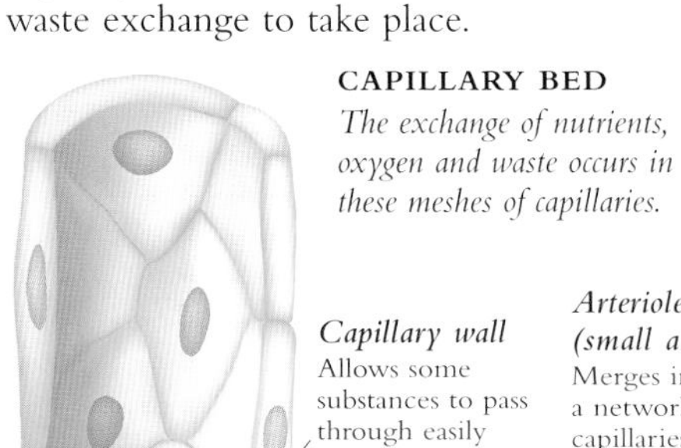

CAPILLARY BED
The exchange of nutrients, oxygen and waste occurs in these meshes of capillaries.

Arteriole (small artery)
Merges into a network of capillaries
Capillary
Venule (small vein)
Capillaries merge into venules

STRUCTURE OF CAPILLARY
The walls of capillaries are only one-cell thick (here each cell and its nucleus are clearly visible), allowing nutrients, oxygen, and waste to pass through easily.

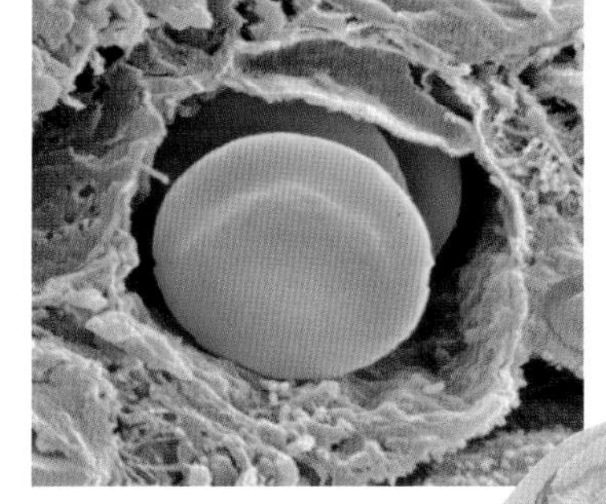

IN A CAPILLARY
In this section through a capillary, individual red blood cells can be seen clearly.

Posterior tibial vein
Posterior tibial artery
Medial plantar artery

BLOOD CIRCULATION
The body's network of vessels and the blood contained in them are the body's transport system. Arteries, veins, and capillaries keep the whole body supplied with blood.

Profile

William Harvey

English physician William Harvey (1578–1657) was the first to prove that blood is pumped around the body in a closed circuit – without leaking or being consumed by organs (as was previously believed). Harvey's discovery led to a new way of treating patients whose lives were at risk as a result of blood loss or illness, by transferring blood into their veins from a healthy donor. His work led the practice of blood transfusions.

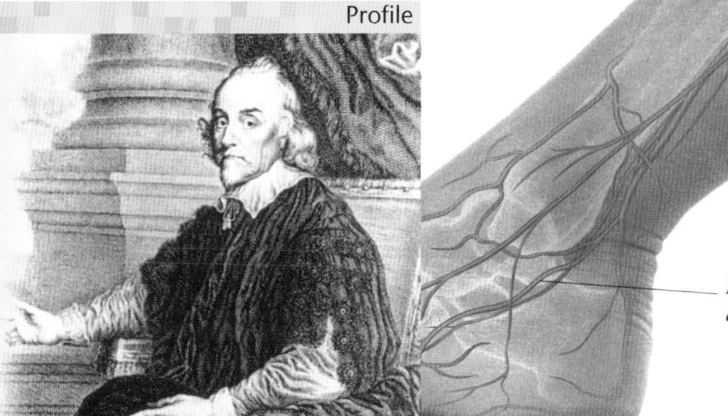

BODY

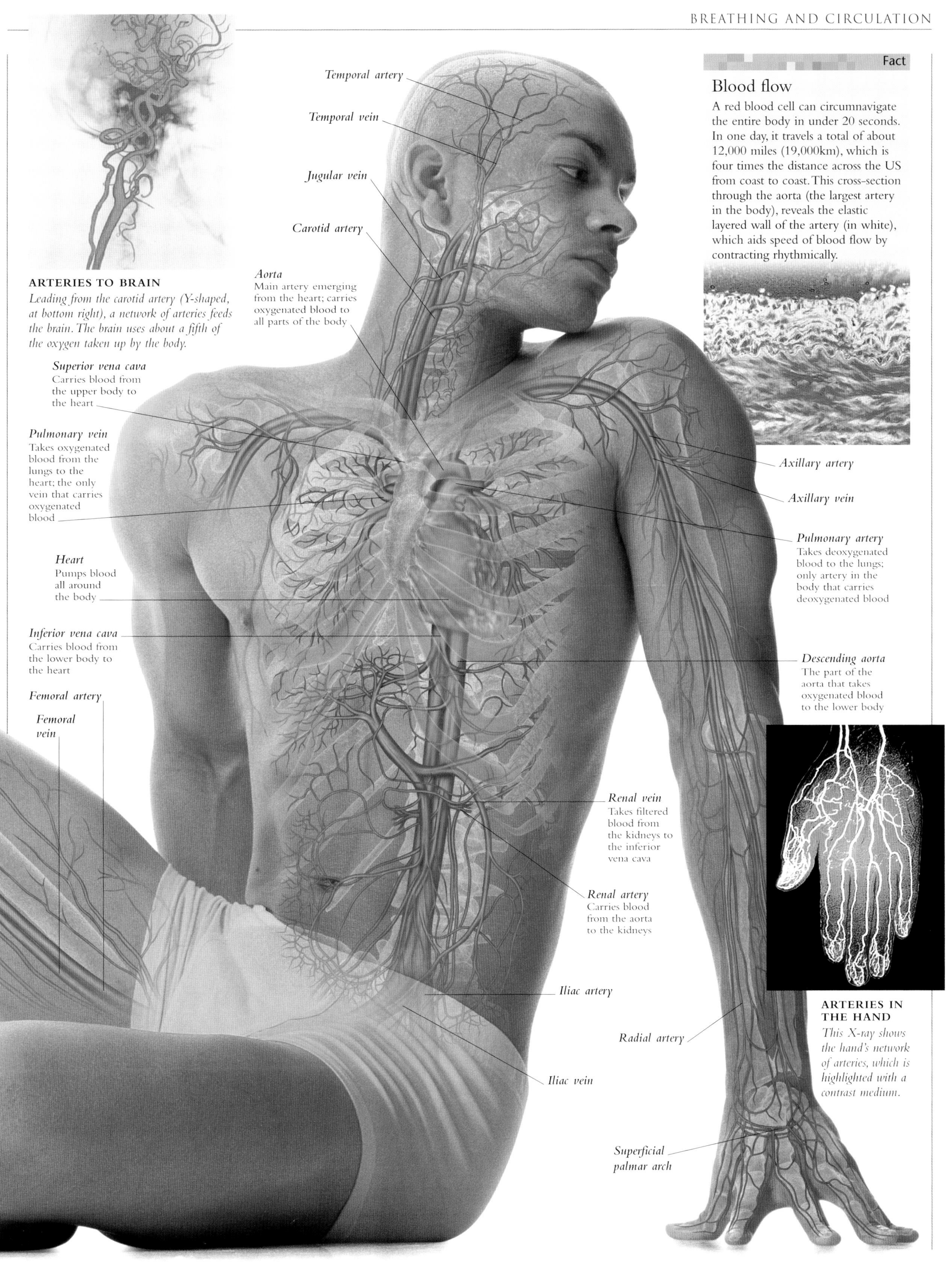

ARTERIES TO BRAIN

Leading from the carotid artery (Y-shaped, at bottom right), a network of arteries feeds the brain. The brain uses about a fifth of the oxygen taken up by the body.

Fact

Blood flow

A red blood cell can circumnavigate the entire body in under 20 seconds. In one day, it travels a total of about 12,000 miles (19,000km), which is four times the distance across the US from coast to coast. This cross-section through the aorta (the largest artery in the body), reveals the elastic layered wall of the artery (in white), which aids speed of blood flow by contracting rhythmically.

ARTERIES IN THE HAND

This X-ray shows the hand's network of arteries, which is highlighted with a contrast medium.

THE HEART

The heart is the body's driving force. This powerful organ, the size of a large clenched fist, sits between the lungs, tilted towards the left side of the body. It works continuously, sending blood – about 5 litres (9pt) a minute – through the lungs and around the body, so that life-giving oxygen reaches every cell. In an average human lifetime, the heart beats over 3 billion times. The heart's special muscle contracts automatically, with no need for instructions from the brain.

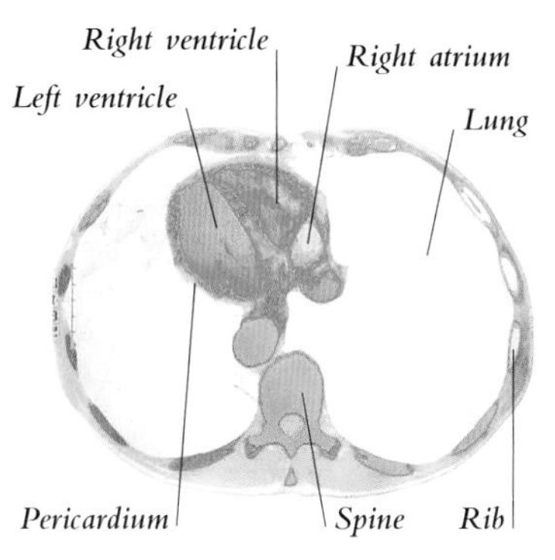

LOCATION OF THE HEART
The heart (seen here from above) sits tucked between the lungs and is enveloped in a thick, double-layered membrane, the pericardium.

THE BODY'S PUMP

When circulating blood flows into the heart, it rushes through a series of inner chambers, pumped from one to another by the contractions of the heart muscle. There are two upper chambers, called atria (a single one is an atrium), and two lower chambers, known as ventricles. The right atrium fills with used blood that has deposited its supply of oxygen around the body. The left atrium collects blood enriched with oxygen picked up in the lungs. When full, the atria contract, squeezing blood into the ventricles below. The ventricles have a harder task than the atria, and for this reason their walls are thicker, especially on the left side of the heart. These chambers contract forcefully enough to push blood out of the heart and back into the blood vessels. Blood from the right ventricle flows through the pulmonary arteries to the lungs. The left ventricle sends oxygen-rich blood around the entire body.

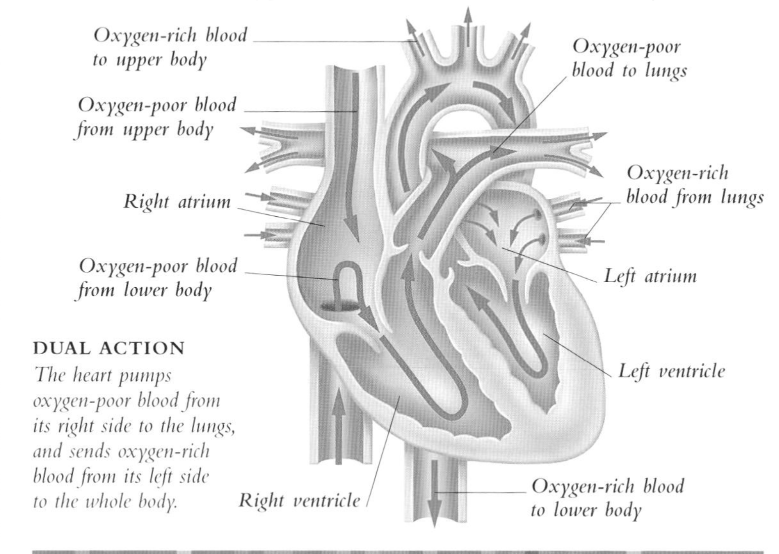

DUAL ACTION
The heart pumps oxygen-poor blood from its right side to the lungs, and sends oxygen-rich blood from its left side to the whole body.

Health

Coronary artery disease

One of the most common causes of heart attacks is narrowing of the coronary arteries that supply the heart with blood (see ringed area in photograph). Coronary artery disease (CAD) is usually due to fatty deposits in the artery wall. The disease is often linked to obesity, a high-fat diet, lack of exercise, smoking, and a family (genetic) disposition. Coronary artery disease occurs more frequently among Western societies.

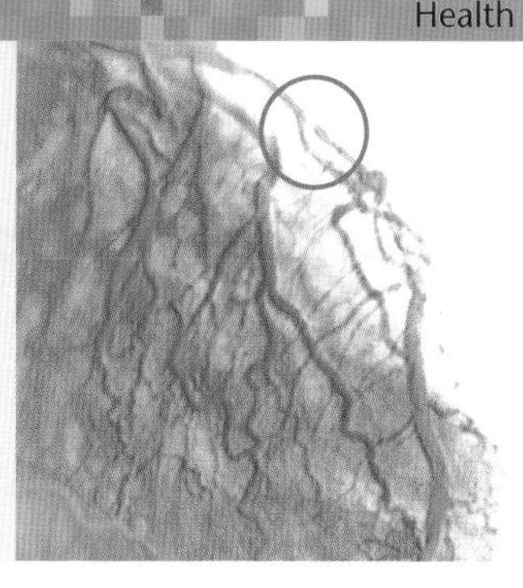

CONTROLLING BLOOD FLOW

Blood flow between the atria and the ventricles is controlled by one-way valves. There are also valves at the openings to the blood vessels leading out of the heart. Each valve consists of two or three cup-shaped flaps of fibrous tissue called cusps, which prevent the blood from flowing backwards. When the atria contract, they push their load of blood against the valves leading into the ventricles, forcing the cusps open. As the ventricles fill and begin to contract, the pressure of blood rises on the other side of the valves and slams the cusps tightly shut. In the same way, blood leaving the ventricles builds up a high pressure that closes the valves at the ventricle exits. The closing of heart valves causes the familiar "lub-dub" sound of the heartbeat that is heard through a stethoscope.

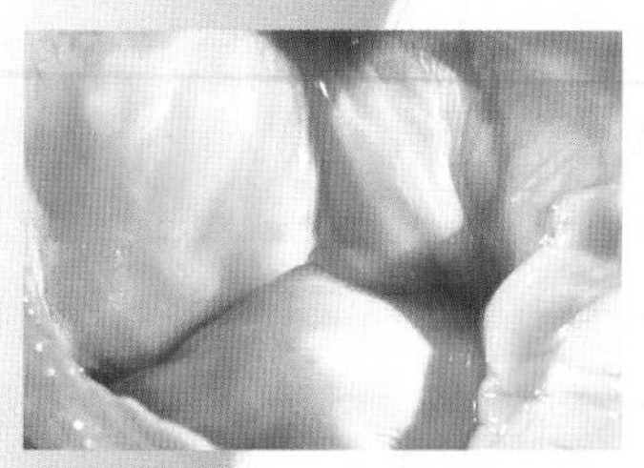

PULMONARY VALVE
The three cup-shaped cusps of the pulmonary valve close tightly together to prevent blood flowing backwards as it leaves the right ventricle.

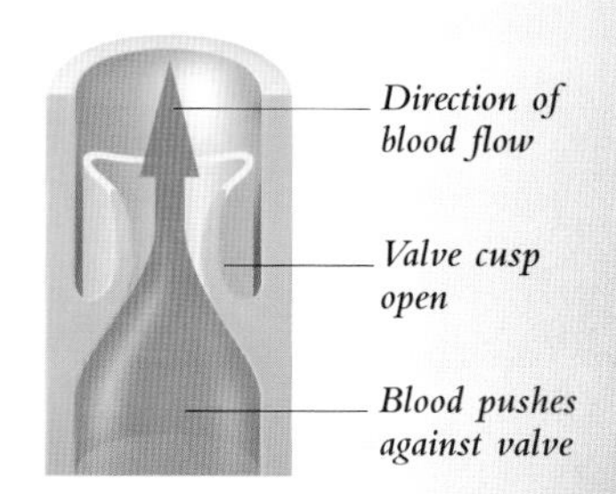

HEART VALVE OPEN
As a heart chamber contracts it pushes blood up against a valve, forcing the cusps to open.

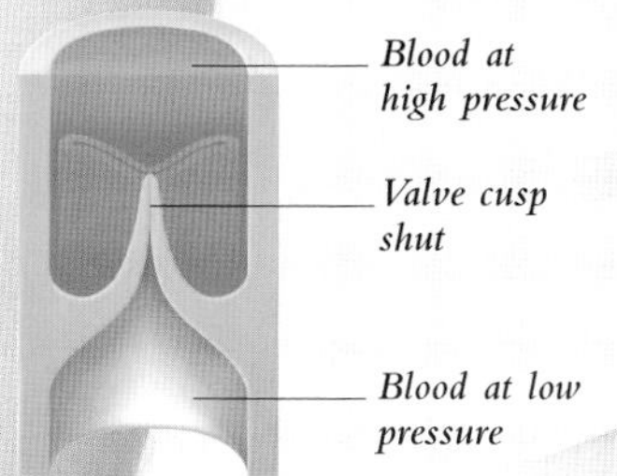

HEART VALVE CLOSED
On the far side of the valve, rising blood pressure slams the cusps shut, preventing backflow.

BLOOD SUPPLY TO THE HEART

Like any other organ in the body, the heart is constantly hungry for a supply of oxygen-rich blood to ensure that it is always functioning efficiently. However, the heart cannot directly absorb any of the blood that it pumps ceaselessly through its chambers. Heart tissue needs its own separate blood supply. To serve this purpose, a network of blood vessels, known as the coronary system, spreads over the surface of the heart. The main blood supply lines serving the heart are the two coronary arteries. These arteries branch from the aorta, which is the largest blood vessel in the body, and subdivide into smaller blood vessels that penetrate the heart muscle. Once oxygen has been delivered to the heart, the coronary veins carry away the used blood and take it back to the heart's right atrium.

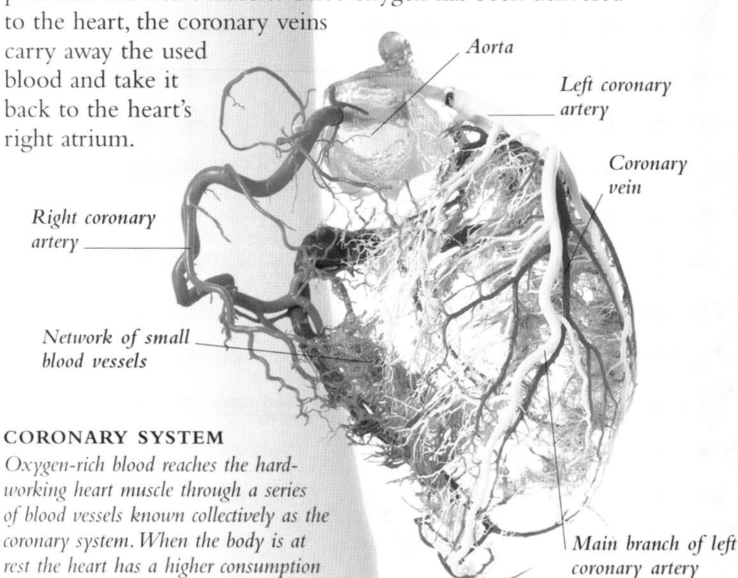

CORONARY SYSTEM
Oxygen-rich blood reaches the hard-working heart muscle through a series of blood vessels known collectively as the coronary system. When the body is at rest the heart has a higher consumption of oxygen than almost any other organ.

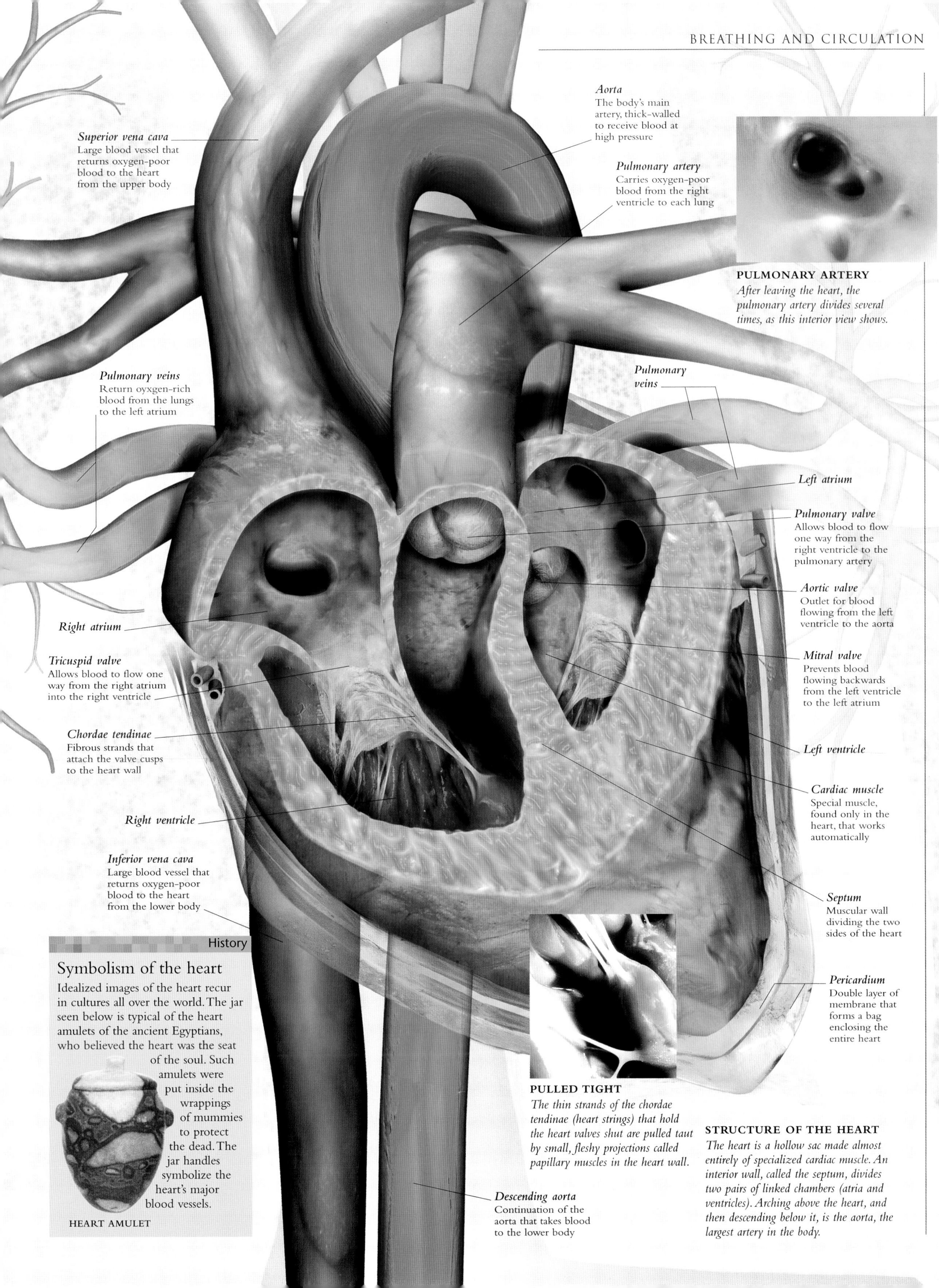

Superior vena cava
Large blood vessel that returns oxygen-poor blood to the heart from the upper body

Aorta
The body's main artery, thick-walled to receive blood at high pressure

Pulmonary artery
Carries oxygen-poor blood from the right ventricle to each lung

PULMONARY ARTERY
After leaving the heart, the pulmonary artery divides several times, as this interior view shows.

Pulmonary veins
Return oxygen-rich blood from the lungs to the left atrium

Pulmonary veins

Left atrium

Pulmonary valve
Allows blood to flow one way from the right ventricle to the pulmonary artery

Aortic valve
Outlet for blood flowing from the left ventricle to the aorta

Right atrium

Tricuspid valve
Allows blood to flow one way from the right atrium into the right ventricle

Mitral valve
Prevents blood flowing backwards from the left ventricle to the left atrium

Chordae tendinae
Fibrous strands that attach the valve cusps to the heart wall

Left ventricle

Cardiac muscle
Special muscle, found only in the heart, that works automatically

Right ventricle

Inferior vena cava
Large blood vessel that returns oxygen-poor blood to the heart from the lower body

Septum
Muscular wall dividing the two sides of the heart

Pericardium
Double layer of membrane that forms a bag enclosing the entire heart

History

Symbolism of the heart

Idealized images of the heart recur in cultures all over the world. The jar seen below is typical of the heart amulets of the ancient Egyptians, who believed the heart was the seat of the soul. Such amulets were put inside the wrappings of mummies to protect the dead. The jar handles symbolize the heart's major blood vessels.

HEART AMULET

PULLED TIGHT
The thin strands of the chordae tendinae (heart strings) that hold the heart valves shut are pulled taut by small, fleshy projections called papillary muscles in the heart wall.

Descending aorta
Continuation of the aorta that takes blood to the lower body

STRUCTURE OF THE HEART
The heart is a hollow sac made almost entirely of specialized cardiac muscle. An interior wall, called the septum, divides two pairs of linked chambers (atria and ventricles). Arching above the heart, and then descending below it, is the aorta, the largest artery in the body.

THE HEART CYCLE

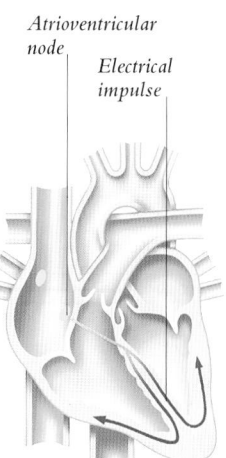

A single pumping action of the heart is called a heartbeat. When a person is at rest, his or her heart beats at a rate of 60–80 beats per minute, but during strenuous exercise this can rise to up to 200 beats per minute. Inside the heart, one-way valves prevent the blood from being pumped in the wrong direction. The characteristic rhythmic "lub-dub" sound of the heart is caused by these heart valves shutting tightly. A heartbeat has three phases. In diastole, the heart relaxes; during atrial systole, the atria (upper chambers) contract; and in ventricular systole, the ventricles (lower chambers) contract. The sinoatrial node (the heart's natural pacemaker) regulates the timing of these phases by sending electrical impulses to the atria and ventricles. Below, electrical activity is shown on an electrocardiogram (ECG) tracing.

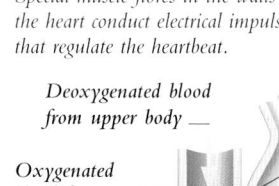

CONDUCTING FIBRES

Special muscle fibres in the walls of the heart conduct electrical impulses that regulate the heartbeat.

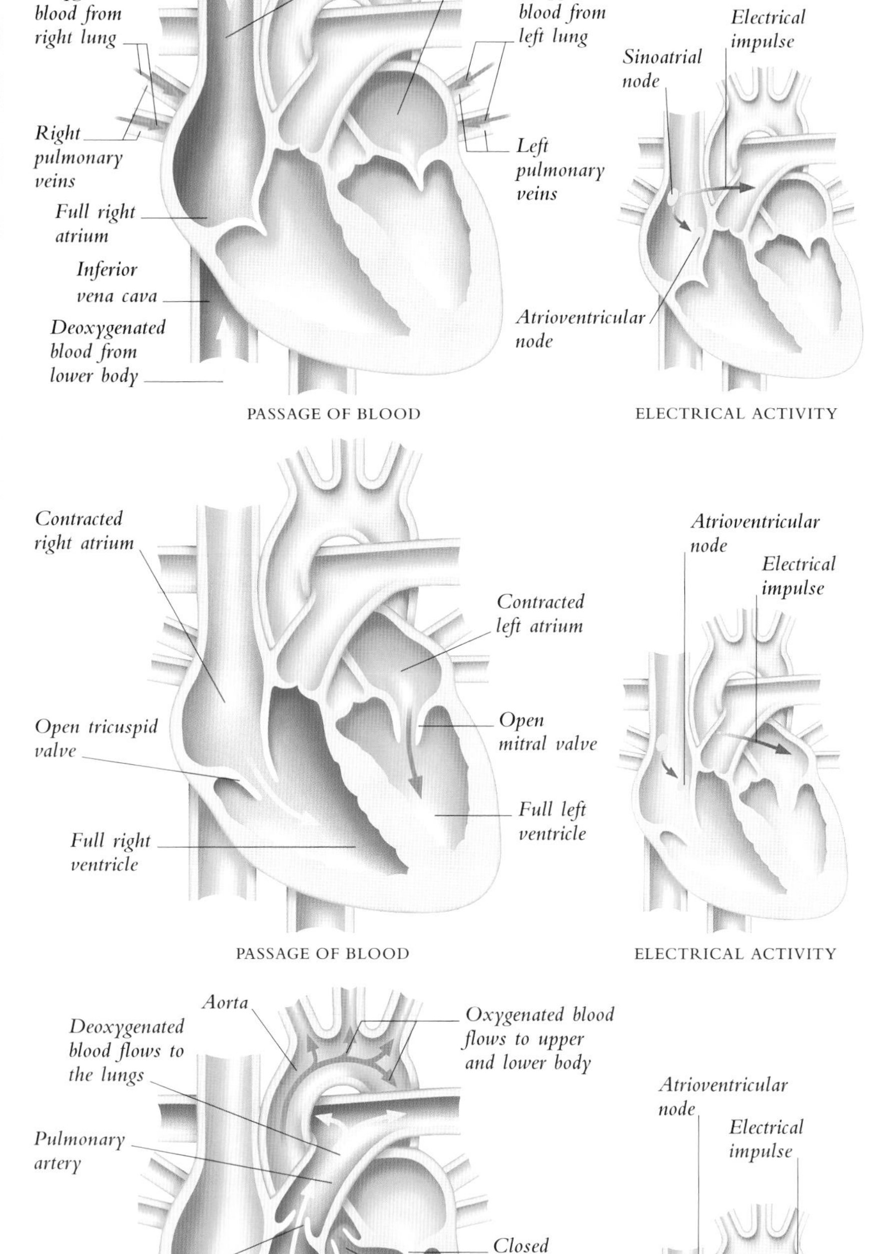

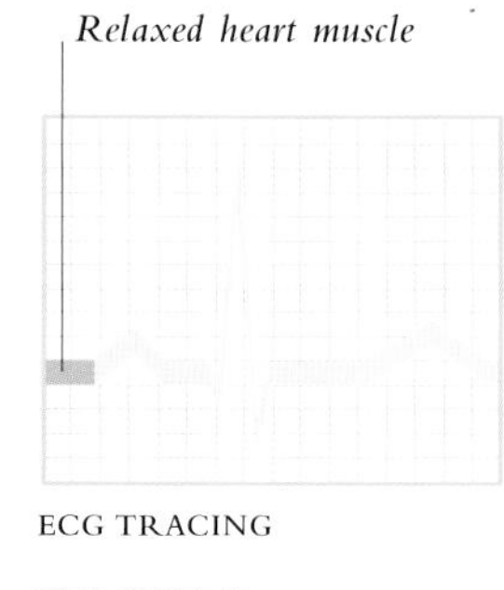

ECG TRACING

DIASTOLE

The heart muscle relaxes and blood flows into the atria from the pulmonary veins and vena cava. Near the end of this phase of the heart cycle, the sinoatrial node emits an electrical impulse.

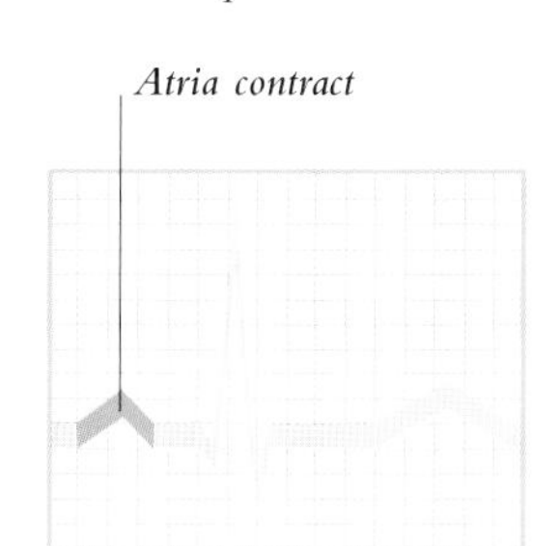

ECG TRACING

ATRIAL SYSTOLE

The electrical impulse from the sinoatrial node spreads through both atria, causing their walls to contract and push blood into the ventricles. The impulse then reaches the atrioventricular node.

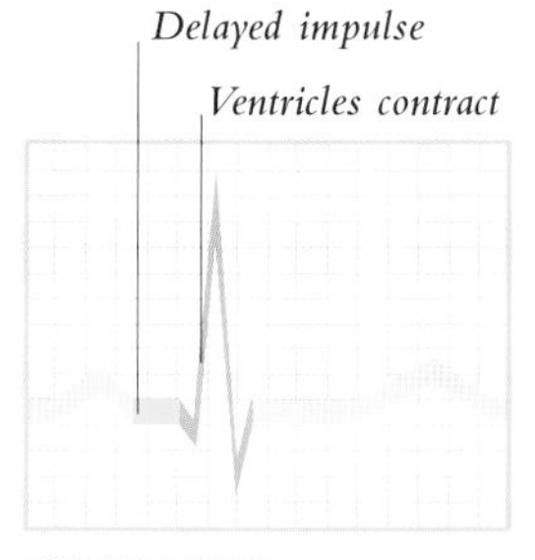

ECG TRACING

VENTRICULAR SYSTOLE

The electrical impulse is delayed at the atrioventricular node. It then spreads through the walls of the ventricles, so that the ventricles contract at the same time, pushing blood into the aorta and pulmonary artery.

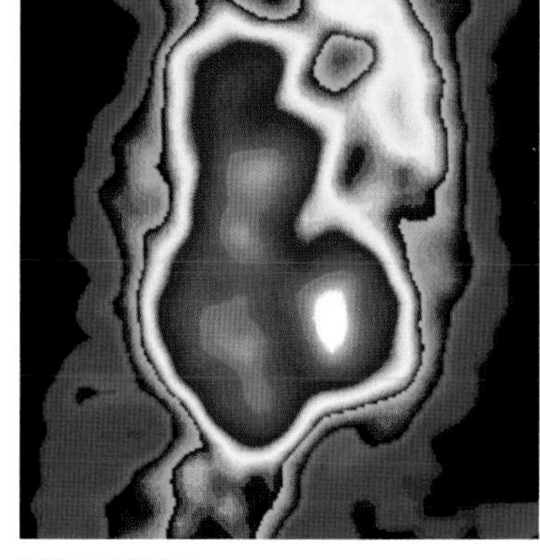

RELAXATION

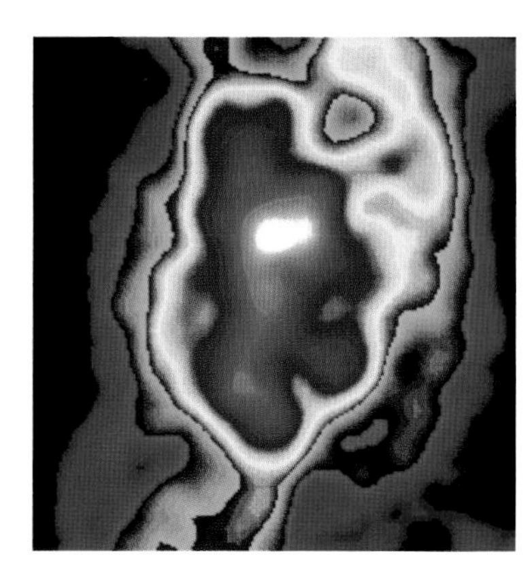

CONTRACTION

DISTRIBUTION OF BLOOD

These two scans of the heart show how the distribution of blood (red) in the heart varies at different stages of the pumping cycle. At top, the heart is relaxed and in the process of filling with blood. Above, the heart is contracted and in the process of squeezing blood out.

Health

Pacemaker

An irregular heartbeat can be treated with a pacemaker that is surgically implanted in the chest. The device is inserted just under the skin in the chest (it is usually visible as a small bulge of skin) and supplies rhythmic electrical impulses along wires to the heart, to keep it beating regularly. An irregular heartbeat may be caused when the sinoatrial node (the area of the heart that initiates heartbeat) malfunctions, or when its impulses are blocked, for example by damage to surrounding tissue.

LOCATION OF A PACEMAKER

This coloured X-ray shows a pacemaker (bottom right), attached by two wires (blue and red) to an enlarged heart.

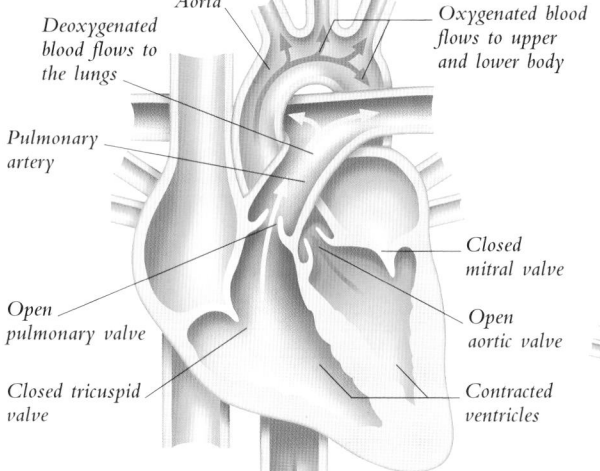

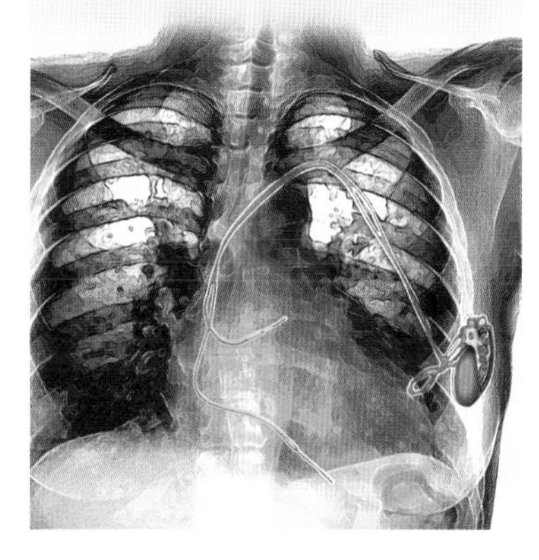

BODY

BLOOD CIRCULATION

Blood circulates in two linked circuits: the pulmonary circulation, which carries blood to the lungs to be oxygenated, and the systemic circulation, which supplies oxygenated blood to the body. Arteries carrying blood from the heart divide into smaller vessels called arterioles and then into capillaries, where oxygen, nutrient, and waste exchange occurs. Capillaries join up to form venules, which in turn join up to form veins that carry blood back to the heart. The portal vein does not return blood to the heart but carries it to the liver (*see* p92). The heart powers both the pulmonary and the systemic circulations, shown in the illustration at right. In the pulmonary circulation, deoxygenated blood (shown in blue) travels to the lungs, where it absorbs oxygen before returning to the heart. This oxygenated blood (shown in red) is pumped around the body in the systemic circulation. Body tissues absorb oxygen, and deoxygenated blood returns to the heart in order to be pumped to the lungs again.

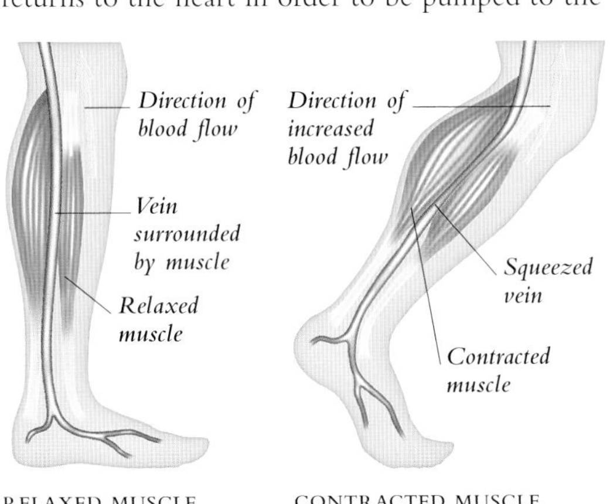

Direction of blood flow
Vein surrounded by muscle
Relaxed muscle
Direction of increased blood flow
Squeezed vein
Contracted muscle
RELAXED MUSCLE
CONTRACTED MUSCLE

VENOUS RETURN

The blood pressure in the veins is about a tenth of that in the arteries. A variety of mechanisms ensure that there is adequate venous return (blood flow back to the heart). Muscles contract and relax as we move, squeezing the veins that pass through them and pushing blood back to the heart.

A DOUBLE CIRCUIT

Here the pulmonary circulation, carrying blood to the lungs, is shown by green arrows and the systemic circulation, supplying oxygen-rich blood to the body, is shown by the yellow arrows.

Network of vessels in upper body

Aorta Carries oxygenated blood to all parts of the body

Pulmonary vein Carries oxygenated blood from the lungs back to the heart

Arteriole

Capillary

Venule

Network of vessels in right lung Blood gives up carbon dioxide and absorbs oxygen in the lung capillaries

Network of vessels in left lung

Pulmonary artery Takes deoxygenated blood to the lungs

Superior vena cava Carries blood from the upper body to the heart

Portal vein Carries blood rich in nutrients from the digestive system to the liver

Inferior vena cava Carries blood from the lower body to the heart

Network of vessels in liver

Network of vessels in digestive system

Network of vessels in lower body

HORMONE ACTION

Various hormones raise or lower blood pressure over a period of several hours and may remain effective for days.

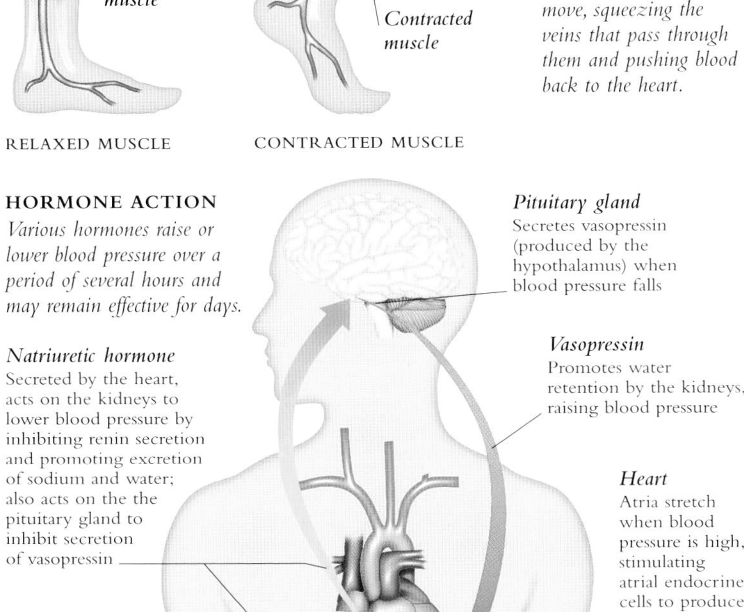

Natriuretic hormone Secreted by the heart, acts on the kidneys to lower blood pressure by inhibiting renin secretion and promoting excretion of sodium and water; also acts on the the pituitary gland to inhibit secretion of vasopressin
Adrenal gland Produces the hormone aldosterone when stimulated by angiotensin
Aldosterone Causes the kidneys to retain salts, increasing fluid in the body and raising blood pressure
Pituitary gland Secretes vasopressin (produced by the hypothalamus) when blood pressure falls
Vasopressin Promotes water retention by the kidneys, raising blood pressure
Heart Atria stretch when blood pressure is high, stimulating atrial endocrine cells to produce natriuretic hormone
Kidney Produces the hormone renin when blood pressure is low
Renin Activates angiotensin in the blood vessels, so that they constrict and raise blood pressure

BLOOD PRESSURE CONTROL

Blood pressure in the arteries must be regulated to ensure that there is always an adequate supply of blood, and therefore oxygen, to the organs. If arterial pressure is too low, not enough blood reaches body tissues. If, on the other hand, the pressure is too high, it may damage blood vessels and organs. Rapid changes in blood pressure (caused by heavy bleeding or a change in posture, for example) trigger responses from the nervous system within seconds. These autonomic nervous responses do not involve the conscious parts of the brain. Longer-term changes in blood pressure (caused by stress, for example) are largely regulated by hormones that affect the volume of fluid excreted by the kidneys. Hormonal responses work over a period of several hours.

BLOOD PRESSURE CYCLE

Arterial pressure is low while the heart fills with blood (diastolic pressure), but rises as the heart pumps blood out (systolic pressure). Pressure is measured in millimetres of mercury (mmHg).

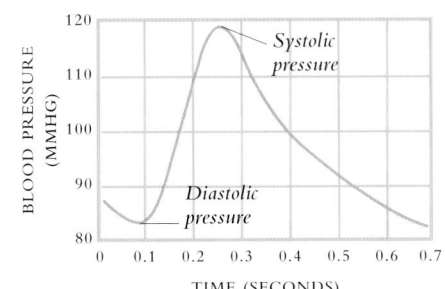

BLOOD PRESSURE (MMHG): 80, 90, 100, 110, 120

TIME (SECONDS): 0, 0.1, 0.2, 0.3, 0.4, 0.5, 0.6, 0.7

Systolic pressure

Diastolic pressure

Fact

Hypertension

Persistent high blood pressure, called hypertension, may damage the arteries and the heart. The condition is most common in middle-aged and elderly people. Genetic factors may contribute, as well as lifestyle factors such as being overweight and drinking excessive amounts of alcohol.

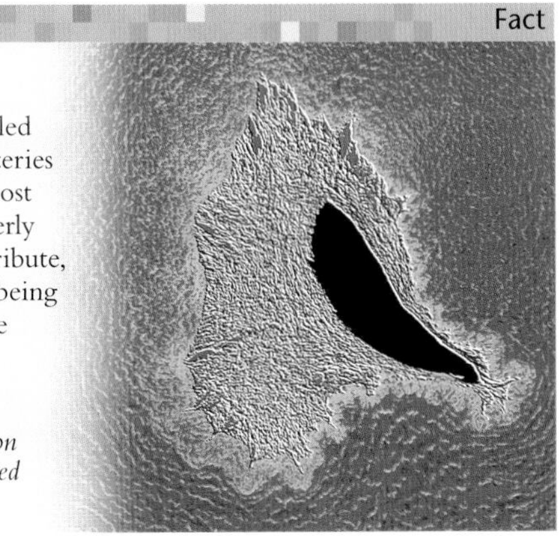

NARROWED ARTERY

A build up of fatty deposits (brown) on the wall of this artery (red) has occurred as a result of hypertension.

BLOOD

The red liquid flowing in our network of arteries and veins has many roles. Blood is a supply line, transporting everything that the body needs to function efficiently. This essential fluid circulates constantly, through the lungs and around the entire body, taking oxygen and nutrients such as sugar, fats, and proteins to all body tissues. Blood also removes toxic wastes produced by cells, and helps to keep the body at the right temperature. Vitally, as part of our natural defence system, blood rapidly delivers disease-fighting cells to areas threatened by dangerous organisms.

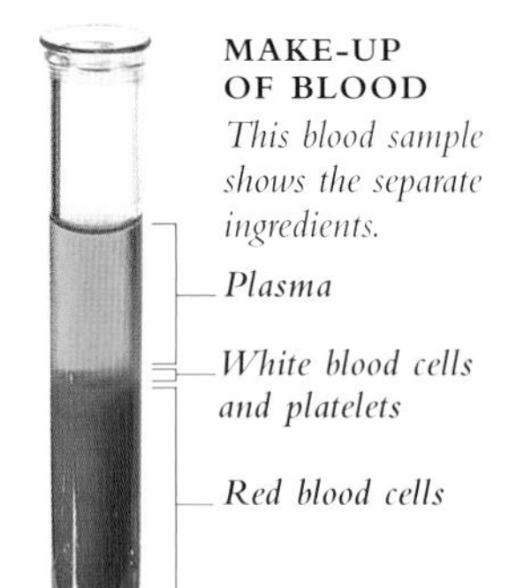

MAKE-UP OF BLOOD
This blood sample shows the separate ingredients.

BLOOD INGREDIENTS

Blood consists of a pale, straw-coloured liquid called plasma, in which float billions of red and white blood cells and platelets. Plasma, which makes up about half the volume of blood, consists mostly of water but contains various dissolved substances, including proteins, salts, and hormones. Red blood cells, by far the most numerous blood cells, are transporters of oxygen and carry away carbon dioxide, the waste product of body cells. The colourless white blood cells are part of the body's inbuilt defence mechanisms. The platelets, irregular-shaped cell fragments, are involved in blood clotting.

Red blood cell
Dimple-shape gives a large surface area for maximum oxygen absorption

OXYGEN TRANSPORTERS

A single drop of blood contains about 5 million red blood cells. These cells carry the red pigment haemoglobin, from which they take their colour. Every haemoglobin molecule carries atoms of iron that attract and pick up oxygen in the lungs. As the red blood cells travel around the body they release their load of oxygen in the tissues. Highly flexible, the cells can squeeze through the tiniest blood vessels to reach every part of the body. Unlike most cells in the body, red blood cells do not have a nucleus.

Neutrophil
The most common type of white blood cell, which targets mainly bacteria

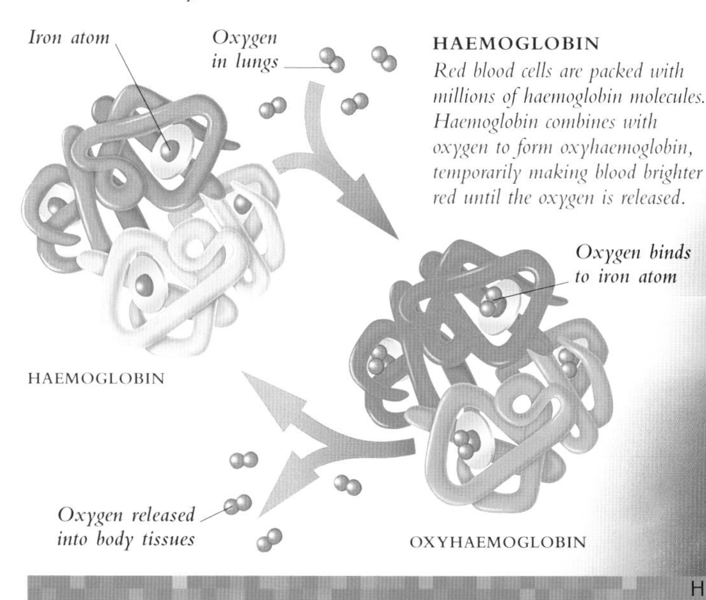

HAEMOGLOBIN
Red blood cells are packed with millions of haemoglobin molecules. Haemoglobin combines with oxygen to form oxyhaemoglobin, temporarily making blood brighter red until the oxygen is released.

Health

Sickle-cell disease

The misshapen red blood cells seen here are caused by the inherited blood disorder sickle cell disease. A defect in the production of haemoglobin, the oxygen-gathering pigment, results in fragile cells that distort into a sickle shape when oxygen levels in the blood are low. These sickle cells may block narrow blood vessels, causing pain and preventing oxygen reaching body tissues. The disease is most often found in African–American people.

IN THE BLOODSTREAM
Various types of cells tumble in plasma, as blood rushes around the body on an endless circuit. Red blood cells, which give blood its colour, predominate. White blood cells and tiny platelets are sparsely scattered in comparison, but they have greater versatility and more active roles.

Blood vessel wall
Elastic structure withstands the pressure of circulating blood

Lymphocyte
One of a group of white blood cells that target specific infections and cancers

Platelet
Clumps with other platelets to seal damaged blood vessels

PROTECTIVE CELLS

Billions of specialized cells circulate in the blood as part of the body's protective mechanism (*see also* Defence and repair, pp96–105). White blood cells of a variety of types have the task of tracking down and destroying harmful organisms. These blood cells migrate through the bloodstream, some engulfing and digesting bacteria and foreign particles, while others target cancer cells and specific infections. White blood cells are larger than red blood cells and have a nucleus. They are classified according to their role and appearance; the main types are neutrophils, eosinophils, lymphocytes, basophils, and monocytes. The other members of the blood's defence forces are platelets. These are not true cells but minute, disc-shaped cell fragments; like red blood cells they have no nucleus. Platelets move into action when a blood vessel is damaged, clumping together to plug the gap and stem bleeding.

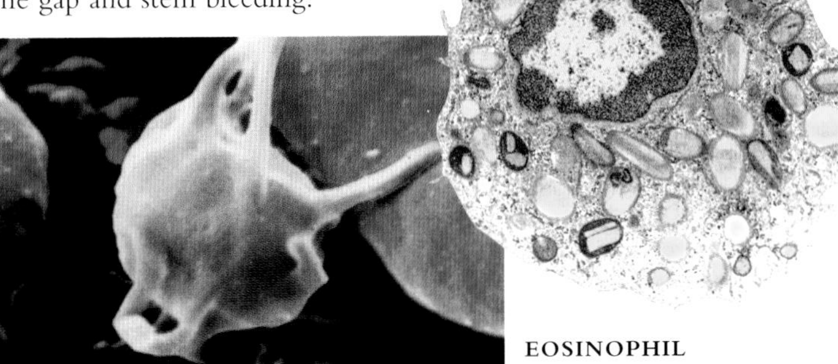

PLATELET
A platelet prepares to take part in the blood-clotting process, forming spiky, adhesive extensions as it latches on and sticks to a red blood cell.

EOSINOPHIL
The type of white blood cell shown in the photograph above contains numerous enzyme granules (in green), that react against foreign organisms such as bacteria.

HOW BLOOD CELLS ARE MADE

All red blood cells and platelets, and most white blood cells, form in the marrow of bones before passing into the bloodstream. The main blood cell production sites are in flat bones such as the breastbone, ribs, shoulder blades, and pelvis. Blood cells have short lives – some white blood cells last only hours and red blood cells are worn out after about 120 days – so constant fresh supplies are needed. Millions of new cells enter the bloodstream every minute. Red blood cells take a few days to mature in the blood before they become fully functioning.

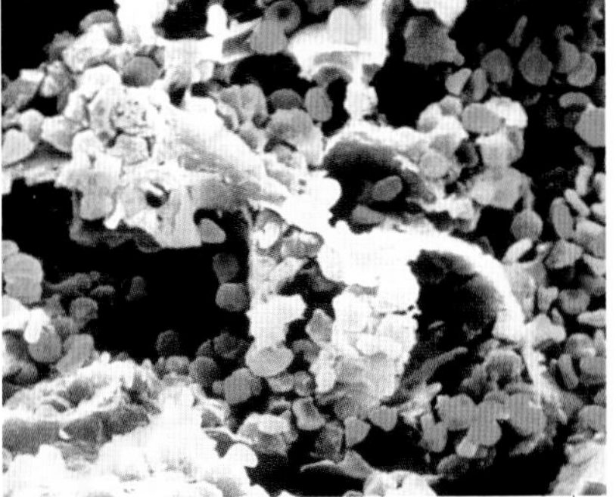

BONE MARROW
Marrow, the soft, fatty substance filling the central cavity of bones, is a factory for the production of new blood cells.

History

Early blood transfusions

The first blood transfusions given to humans were carried out in the 17th century, long before the different blood groups were understood. Using a sheep as a blood donor, as shown in the illustration below, was attempted by a pioneering doctor in 1667. This particular patient survived, but such experiments were more likely to prove fatal.

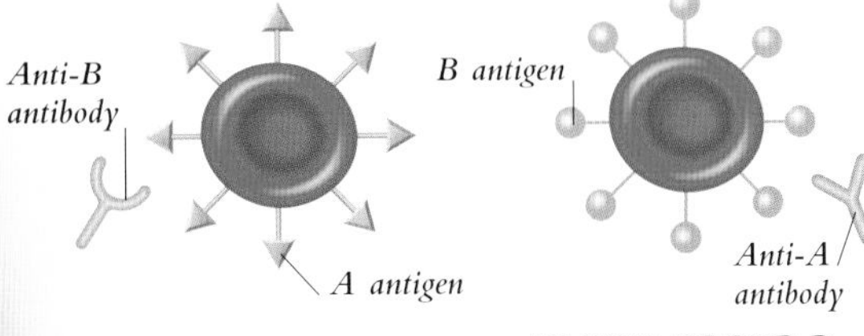

BLOOD GROUP A
This blood type has A antigens on the red blood cells and anti-B antibodies in the plasma.

BLOOD GROUP B
Group B has B antigens on the cells and anti-A antibodies in the plasma.

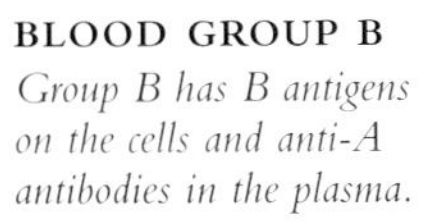

BLOOD GROUP AB
The rare AB group has A and B antigens on the cells and no antibodies in the plasma.

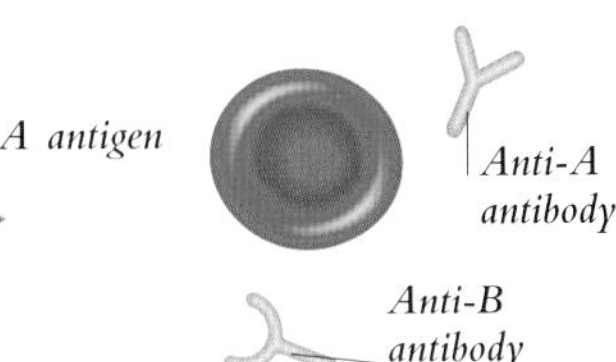

BLOOD GROUP O
O blood has no antigens and contains anti-A and anti-B antibodies.

BLOOD GROUPS

Each person belongs to a blood group. There are many types of blood groupings, of which the best known is the ABO system. This system identifies four groups – A, B, AB, and O – by markers, called antigens, found on the surface of red blood cells. The body's immune system uses antigens to recognize the difference between its own cells and foreign cells. During a blood transfusion, a person must be given blood with the correct antigens. Otherwise, the immune system sees the new red blood cells as invaders and attacks them. In each blood group, proteins in the plasma, antibodies, stick to foreign blood cells, marking them for attack. Another method of blood typing is to identify the rhesus (Rh) antigen, found on the red blood cells of 85 per cent of people.

TAKING IN FUEL

The sight and smell of food triggers the start of the digestive process by stimulating the flow of saliva in the mouth and the secretion of gastric juices in the stomach.